Immune-based Cancer Treatment

The T lymphocyte Response

Immune-based Cancer Treatment

The T lymphocyte Response

Michael A. Alexander
Widener University
Chester, PA
USA

CRC Press
Taylor & Francis Group
an **informa** business
www.crcpress.com

6000 Broken Sound Parkway, NW
Suite 300, Boca Raton, FL 33487
270 Madison Avenue
New York, NY 10016
2 Park Square, Milton Park
Abingdon, Oxon OX 14 4RN, UK

Science Publishers
Enfield, New Hampshire

Published by Science Publishers, P.O. Box 699, Enfield, NH 03748, USA
An imprint of Edenbridge Ltd., British Channel Islands

E-mail: info@scipub.net Website: www.scipub.net

Marketed and distributed by:

CRC Press
Taylor & Francis Group
an informa business
www.crcpress.com

6000 Broken Sound Parkway, NW
Suite 300, Boca Raton, FL 33487
270 Madison Avenue
New York, NY 10016
2 Park Square, Milton Park
Abingdon, Oxon OX 14 4RN, UK

ISBN 978-1-57808-703-7

Library of Congress Cataloging-in-Publication Data

Alexander, Michael A. (Michael Allen), 1953-
Immune-based cancer treatment : the T lymphocyte response / Michael A.
Alexander.
p. ; cm.
Includes bibliographical references and index.
ISBN 978-1-57808-703-7 (hardcover)
1. Cancer--Immunological aspects. 2. T cells. 3. Cancer--Immunotherapy. I. Title.
[DNLM: 1. Neoplasms--immunology. 2. Neoplasms--therapy. 3. Immunotherapy. 4. T-Lymphocytes--immunology. QZ 266]
RC268.3.A435 2011
616.99'4061--dc22

2010046622

Printed in the United States of America

Foreword

Economists have estimated that a 1% reduction in cancer mortality (deaths) results in the addition of more than US$ 500 billion to the national wealth. Despite this, there is a cancer-funding crisis. For example, the government's 2007 budget for cancer research through the National Cancer Institute (NCI) wanted to appropriate US$ 4.75 billion. This was $40 million less than in 2006 and $71 million less than in 2005. Also, in 2007, the NCI's Clinical Trials Cooperative Group Program was ordered by the government to cut 60% of the proposed new clinical trials. As a result, 2,000 to 3,000 patients were left out of new trials. This was a money-saving maneuver that cost lives.

Most people have no idea how many people die from cancer. They would be surprised to find that there are about 2,500 deaths a day. Also, those who die represent a cross-section of our population. Not all are old; some are very young, even though the older you get, the more susceptible you are to cancer because it generally takes time to develop. As a matter of fact, because of the baby boom, 20% of the U.S. population will be over 65 in 2030. For this reason the deaths per day will increase almost exponentially (times ten) each day within the next 5–10 years if things remain the same. Race inequities and poverty aggravate the situation because some people wait until it is too late to see a doctor and die shortly after their diagnosis. This disease takes some of the most vibrant, productive, and important people we have.

If you are 95 years old and die of cancer it's probably not that unusual. However, if you are between 5 and 95 years of age it's another matter. We live as though we will live forever, but the reality is that at any moment any one of us could be diagnosed with cancer. Then we may have to endure cancer treatments that could be painful and we could still die. Getting cancer is something quite indescribable and only fully understood when experienced first hand. The physical part of the experience can be either benign or beyond comprehension. Surely, the most insidious part of the experience is the emotional and physiological aspects. As you can imagine, physical problems (such as pain) subside and eventually go away, but the emotional and psychological memories of the experience remain, much like an unseen surgical scar. The fact is that we do not live forever, but we may forever have to live with having gotten cancer.

Acknowledgements

Since this is the first of what I hope is a series of books that I will publish in the future, I want first to acknowledge and thank the person most responsible for my ability to do such a thing, my mother, Patricia K. Alexander. Without her continual support in my education (academic and life), thinking process, and awareness, none of this would be possible. For someone who does not have a higher education, my mother is one of the most intelligent people I know. Aside from being listed in *Who's Who in America*, she moved the Philadelphia Social Security Administration's computing bureau from key-punch cards to large computer servers. As the Bureau Chief of that department she became the highest-paid African American government employee during the 1960s. I inherited only a fraction of that intellect, yet I was able to be the first one in the family to get a Ph.D. and go on to become an author. This one is for her.

Next, I thank my brother (Kenneth) and my sister (Laura) for their support down through the years. Another person to mention is my best friend of 40 years, Harold Johnson, who has helped me immensely with those life problems we all face at one point or another. Without knowing him, I would probably not be the person I am today. I am most grateful to my friend and mentor Dr. DuPont Guerry, IV, MD. I've known DuPont for 30 years and from the moment I joined his laboratory at the University of Pennsylvania until now he has been very supportive of my efforts to be a cancer researcher. He allowed me to obtain my Ph.D. at Penn as a full-time graduate student while conducting basic research into the human T cell response to melanoma, which is what my Ph.D. thesis was about. DuPont has taught me first hand not only about compassion for others through his work as an oncologist, but also how to be organized in such a way as to be effective at what I do. For this I am eternally grateful.

This book was written at two sites in Chester City, Delaware County, Pennsylvania. The first site was the Pennsylvania Career Link (Unemployment) Office, which is where the mechanical aspects of the book took shape. I would like to thank Aretha, John, Walter, Ed, and Dianne for being not only patient with me, but also kind and courteous on a daily basis. The actual conceptualization, writing, and editing were performed at the Wolfgram Memorial Library of Widener University. Here is where I spent 10 hours a day reading scientific articles and developing chapter ideas. I used the library's cancer reference books in order to find easily understandable terminology for some of the most complex descriptions I had to write. All in all, it was a wonderful experience being in this library because all the librarians and staff were very kind and helpful. Thanks, Fran. Partly because of my experiences at the University and the fact that I was becoming a writer, I enrolled in a master's program in Liberal Arts, which I hope will improve my writing skills. I would like to thank Lucy and Harry. Lastly, I would like to thank Daniel Russo, newspaper correspondent, who helped me make the adjustment from research writer to book writer. His guidance in this matter was invaluable because he could see the desire in me to successfully make this transition and for that I thank him.

Contents

Introduction

Cancer is one of the most difficult health problems our society faces. Cancer research is a multi-billion dollar endeavor that encompasses the intellectual efforts of scientists around the world. Yet that effort, as of this moment, has not defeated one of the major diseases of our time. Yes, there has been progress, but the fact is that if you get diagnosed with some forms of cancer the likelihood is that you will die from it. This effort to combat cancer is an ongoing process that is sensitive to the economic and ethical practices of our society. In many ways it is similar to a war. This war is waged on many fronts and in many countries. Life for cancer-free people is quite different from the lives of people diagnosed with this disease. For this reason the effort to explain the relationship between our bodies (particularly our T cell immune systems) and cancer is dedicated to them. The estimate of the number of dead and the survivors of cancer just since the 1970s is staggering, but those left behind should know the world is trying to help.

This book is intended to inform individuals about the intricacies of the T cell immune system and how it fights cancer. It's written so that people who are not immunologists can understand the concepts that demonstrate how this system works. The chapters are organized so that they gradually impart information and draw the reader into the subject. These chapters are about how our T cell immune system deals with the major aspects of cancer. Some repetitive sections are left in each chapter to allow the reader to understand how this information (such as autoimmunity) is crucial to the way our T cells attack a newly formed cancer.

The T cell immunological response to cancer is a complicated series of events that occurs without our even knowing it. When that response cannot destroy the cancer, the cancer not only becomes noticeable, it also becomes a major health problem. As you will see from this text, there has been a steady evolution in the development of treatments for cancer. Early on, these treatments were purely chemically based (chemotherapy), but later a deeper understanding of how our immune system works led to an inevitable conclusion: that the immune cells (T cells) in our bodies are quite capable of curing us of our cancer.

For many reasons, this is often not the case for people diagnosed with cancer. The reasons why T cells fail to remove a cancer are becoming clear from our increasing knowledge on the subject of T cell immunology. This understanding has brought about the development of another treatment for cancer, called immunotherapy. Being a relatively new treatment, it has yet to have the impact on the cure rate for cancer that immunologists (like me) had hoped for. However, results of clinical trials using these new methods of treatment are very encouraging. In fact, some of these immune-based techniques are being used to supplement traditional chemotherapeutic treatments. Despite this, many people do not understand immunotherapy or, for that matter, how the T cell immune system works.

In part, this book is intended to do just that. Its goal is to translate a highly scientific and technical subject into language that is self-explanatory and relatively easy to understand. Understanding the elements involved in the development of immunotherapeutic treatments is potentially very important to many people. Current and future cancer patients may use this information to understand this type of treatment and decide whether immunotherapy is right for them. Another interested group would be college and medical school students who wish to pursue a career in T cell tumor immunology. In addition, some physicians may want to understand the basic elements of immunotherapy. For these reasons, it is important for me to make this information available to everyone. It should be considered a starting point, but be warned that in addition to the science, there are comments about our society and cancer research. Detailed information on any aspect of the areas covered in this book can be found on the Internet.

Most people know next to nothing about immunity to cancer. Recently I was talking to a college student. He had seen me in the library quite frequently. He asked me why I was there all day, every day. I told him I was writing a book about immunity to cancer. He was surprised because as far as he knew there was no immune response to cancer. Then I asked him,

"Isn't a cancerous growth in your body a foreign entity?"

He thought about it for a minute and said, "Yes, it is."

Then I asked him, "Isn't our immune system supposed to protect us from foreign entities like bacteria and viruses?"

He said, "Sure."

I said, "Well our immune system protects us from cancer in the same way."

He was clearly perplexed. I think most people when they hear of someone getting cancer do not think of the body fighting it. I went on to explain that almost all of us in our lifetime get cancer, but the T cell immune system destroys the cancer cells at a very early stage. This happens at such an early stage that it is almost undetectable. When the immune system cannot do this the cancer causes us to feel ill and even at that point we don't know it's cancer until the diagnosis. At this point I had got his attention. I had previously explained my credentials (Ph.D. from Penn) and my experience (30 years of cancer research) so he had no problem believing what I was telling him.

I went on to tell this student that there were several factors that contributed to the development of cancer. The first was a predisposition to getting cancer, when a gene that causes cancer is passed down through families. This happens in some cases of breast and ovarian cancer. The other factor has to do with our environment, where cancer-causing agents such as asbestos, radon gas, and benzene from exhaust fumes induce cancerous cells. Or it can be the combination of these factors. Cancer is foreign, but because it develops from our own cells the immune system may not see it in time to eliminate it. In other words, a part of the immune system thinks a developing cancer cell is a normal cell and actually protects it from being destroyed. He did not quite understand that.

So I told him, "Look at it this way. A woman is pregnant. The cells in the fetus are half hers and half the father's. Her immune system, which would normally see the father's cells and destroy them, has built-in safeguards to protect that fetus until the baby is born. Sometimes these safeguards come into play when a cancer develops, but by the time they go away and the immune cells (T cells) are ready to attack the cancer, it is too late. By that time the cancer cells have vastly outnumbered the T cells and they have spread throughout the body." I think he was awestruck at this last statement and he had clearly learned something new.

This type of thing has often happened to me. One thing I learned during these explanatory encounters is that people can't learn something that they don't understand from the beginning. Cancer research and immunity for many people is like a foreign language. Not only is the technology of cancer complicated, the language of cancer is, too. Cancer doctors (oncologists) have an increasingly difficult time explaining to a patient what type of cancer they have and how it is going to be treated. Therefore, as I started this endeavor I had to keep in mind the fact that I had to work hard at simplification. I am by no means trying to insult anyone's intelligence, but this was a very difficult task. Some of the material in this book will be challenging for some readers, but some of it was very challenging for me as well. I had to learn information that was entirely new to me and then translate it. I had to change complex terms into language that was understandable to reasonably educated people and in some cases repeat the definition several times throughout the text. So you, the reader, and I share one thing in common and that is the challenge of understanding the many aspects of the T cell immune response to cancer.

Before you start to read these chapters, I'd like to give you an idea of what's in the book. The first chapter introduces the main types of immune cells in the form of a question about which of these cells control the anti-cancer T cell response. Chapter 2 describes antigen presentation, which is a process essential to the T cell response to cancer. Chapter 3 is a basic view of the interactions between T cells and cancer cells during the anti-cancer T cell response. Chapter 4 looks at the advent and new uses of anti-cancer vaccines, which can instruct the T cells of the immune system to attack a cancer. Chapter 5 introduces anti-cancer immunotherapy, which is a promising treatment because it uses the patient's own T cell immune system to attack the cancer. Chapter 6 describes immunotherapy for leukemia, which is a unique form of cancer that, in some instances, is difficult to treat because of the nature of the cancer cells. It describes how scientists are trying to harness the power of the T cell immune system to fight this type of blood-borne cancer. Chapter 7 introduces a less commonly known problem, that cancer cells can induce autoimmune (anti-self) diseases that can be just as deadly as the cancer itself. Chapter 8 describes the relatively new concept of cancer stem cells, which cause major complications during the anti-cancer T cell response.

Chapter 9 was the most difficult to write because it is based on biochemistry, and my exposure to biochemistry was limited to the purification of cell surface proteins. But I think I managed to describe what translational immunology is, and why it is an important field of study. The next several chapters highlight how translational immunology is being used to treat cancer. In that vein, Chapter 10 highlights the dendritic cell, which is a specialized antigen-presenting cell and plays a pivotal role in directing immune cells before, during, and after the anti-cancer T cell response. Chapter 11 deals with one of the most fascinating aspects of the anti-cancer T cell response, the activity of regulatory T cells or Tregs. Tregs cell can prevent anti-cancer T cells from destroying the cancer because they view the cancer cells as self and protect them from destruction by the anti-cancer T cell immune system. Chapter 12 describes the NK cell. It is not a T cell, but it plays a very important role in the T cell immune response to cancer.

The next few chapters describe a larger realm of events faced by the T cell immune system as it confronts the complexities of a growing cancer. Chapter 13 describes how cancer cells interfere with the T cell response directed at them, before, during, and after a response that would destroy them. Cancer cells do this in both direct and indirect ways, but in the end make it very difficult for T cells to eliminate them. Chapter 14, which is rather long, has to do with the T cell immune response to cancer cells derived from the neural crest. The two highlighted cancers are the brain tumor glioblastoma multiforme and the skin cancer melanoma. Melanoma is a special type of cancer for me because I spent 15 years studying the human T cell response to both its primary (localized) form and its metastatic (widespread) form. There are special problems associated with the T cell response to these cancers because they are derived from neural tissue. The subject of Chapter 15 came to me out of the blue: that our T cell immune systems have holes in them. These holes mean that some of us are unable to respond to quite common immunological threats because of the genetic design of our immune systems. I hope you will find this chapter as interesting to read as I found it to write. Finally, Chapter 16 contains a summary of each chapter and concluding remarks and gives readers a chance to think about what they've just read. The summary can be used as a guide of sorts. Some readers may want to go there after looking at the table of contents and just choose which chapter(s) to read on the basis of the summary.

My hope is that this book will inspire people to take an interest in how our immune system can defeat cancer. The future successful treatment of cancer will certainly involve the manipulation of the T cell immune system. Our T cell immune system can destroy the cancer and leave behind an immunological memory that will prevent the disease from coming back. For it to do that, more information like what is presented here will have to be uncovered by scientists, physicians, and laboratory workers. Discoveries are based on information scientists have gathered through years of research into how our T cells function. Today's new ideas lead to tomorrow's cure for cancer. I invite you, the reader, to explore this new frontier in cancer research.

Chapter 1

Which Cells Control the Anti-cancer T cell Immune Response?

Immunologists down the years have asked which cells control the anti-cancer T cell immune response. The answer is quite perplexing. The response to cancer is probably the most complicated function of the immune system. An army of immune cells and the tumor cell itself are involved. Immunologists have not completely determined the role of these cells in eradicating the tumor. There is a "pecking order" of immune cells involved, but which one or ones can scientists rely on to do what we all want? Namely, cure us of the cancer and never let it return.

Our investigation will delve into this problem. We will examine each cell that has a major role in the T cell response to cancer. We will examine the particular attributes a cell possesses that enable it to positively or negatively affect the anti-cancer T cell response. It is important to understand how each cell involved in the T cell anti-cancer response functions because those are the very cells scientists are going to manipulate to treat cancer. That is what immunotherapy for cancer is about. It channels the function of immune cells toward the inevitable destruction of the cancer. In the future, cancer treatment will be quite different. It should be much less painful and have fewer side effects because the elements used to treat it will come from our own bodies. These are the cells of the immune system that naturally protect us from cancer as well as many other infectious diseases and organisms.

In the immune system, you may have one type of immune cell, but because of the expression of one or more genes in that cell, it behaves differently. In other words, it's like a room full of people all wearing identical clothes, except that groups of them are wearing the same hats. In this case, the same hat denotes a specialized function that that cell performs. This will become clear as these cells are described. Let us narrow down the number of cell types to, say, 12. That seems like a lot of different cells, but actually it is not. They all have different functions. By the way, the 12th cell is the cancer cell. The cancer (tumor) cell actually is a very powerful modulator of the immune response against it. The stem cell nature of a tumor allows it to change and attack the immune system in very clever ways. This ability allows it to sometimes flourish in the face of an immune response that would not only destroy it, but also leave behind memory T cells that would quickly eradicate the tumor if it came back. I am also purposely leaving out the B lymphocytes, although they too can influence the anti-cancer response. This book is about the T cell response to cancer cells and B cells are a separate issue more appropriately dealt with on its own.

We will try to simplify a very complex system involving the relationships these cells have with each other and the cancer cell. These relationships at times may seem redundant. That is an inherent quality of the immune system. The redundancy, or having more than one cell do the same job, is a safeguard. It makes sure we are protected from cancer as well as a world of infectious agents. The T cell immune system is a very intricate natural machine designed to operate on its own. It is only now that we know enough about the T cell immune system to manipulate its function to treat diseases like cancer.

The T cells

There are seven T cells, but only two types. One type is broken down into five functional groups and the other into two, but all of these numbers are in flux. So by the time you read this, another type of T cell may be discovered. Each of these seven T cells is a distinct entity. Let's start with the types and work our way down. The first type is the CD4 T cell and the second is the CD8 T cell. CD4 T cells are generally called helper T cells and CD8 T cells are referred to as cytotoxic T cells. These two cell types serve most of the reconnaissance and enforcement functions of the cellular immune system. Without them there is no immune system. The memory T cells from both of these types are tremendous responders. As a group they are critical to our survival.

The five CD4 T cells

The CD4 T cell is probably the most talented immune cell in the body. It is the first responder and could be considered a director. The CD4 T cell arrives on the scene and, depending on what it sees, can communicate immediately with at least five different cell types. It can direct these cells to perform specialized functions, such as killing cancer cells and virus-infected cells, eating cellular debris, or forming a barrier around an infected site. One of these cells can turn up at an infectious or cancerous event and turn into one of five distinct cells. Once that happens, each of these cells functions in such as way as to separate it from the parent cell, until there is no similarity between it and the parent cell. This phenomenon presents problems for immunologists trying to find out what happened just before a measurable immune response could be detected. That is why this first response is so critical, because to some extent the immune response is predetermined in such a way as to make it almost impossible to change.

The first of these five CD4 T cells is called the type 1 helper CD4 T cell. These cells are created from a new or naive CD4 T cells. They turn into these cell types depending on the type of stimulation they receive. This will be discussed in detail as antigen-presenting cells (APCs) are examined. The type 1 CD4 T cell, when stimulated, releases bioactive proteins called lymphokines. As a rule, CD4 T cells can release 10 to 15 different lymphokines. With each different type of stimulation, the cell releases combinations of lymphokines that instruct other cells to perform specific functions. The type 1 cell releases a combination of lymphokines that activate killer T cells to attack a specific target, for instance, a virus-infected cell or a cancer. Rapid killing of virus-infected cells stops the infection before it can spread. Killing a cancer cell immediately prevents it from developing the ability to move (metastasize) throughout the body.

The second of these five CD4 T cells is the type 2 helper CD4 T cell. It releases a combination of lymphokines that cause antibodies to be produced by B cells against a specific target. This happens during a bacterial infection. The antibody-coated bacterial cells are quickly ingested by macrophages because they have receptors for the end of the antibody not attached to the bacteria. The ingestion process actively induces a cascade of events called antigen presentation. This process will be discussed at length later when macrophages are described. The type 2 CD4 T cell can also, through the elaboration of lymphokines, prevent the development of killer T cells. Sometimes the exclusion of killer T cells is not a good thing, but that's how the system works.

The next two CD4 T cells, unlike the first two, can induce a negative immune response. On the face of it, we would wonder why there should be such cells. The reason has to do with the efficiency or over-efficiency of the T cell immune system. There is one word to briefly describe this situation: autoimmunity. Many of the diseases we suffer from are autoimmune diseases or the result of inappropriate autoimmune reactions. Diabetes, lupus, arthritis, and multiple sclerosis are just a few examples. The common theme is that the immune system, using either cells or antibodies, attacks normal tissues. The reasons for that type of self-attack are numerous and sometimes unintentional, but there has to be a system in place for the immune system itself to regulate or stop it.

This is where the regulatory arm of the T cell immune system comes in. Like many other functions, the regulatory arm falls under the auspices of the CD4 T cell. T regulatory cells are unique. Some are created from the CD4 T cell pool, depending on the stimulus encountered. Some are created in the bone marrow, pass through the thymus (checkpoint for all T cells), and then are released into the blood to be called upon if needed. One of the two types of these cells (Tregs) has a specialized gene expressed called Fox 3 and expresses a high-affinity receptor (CD25) for the lymphokine interleukin-2 (IL-2). This cell, like many other CD4 T cells, exists in a fluid state. It can be changed from a Treg into a responding CD4 T cell if needed. Conversely, Tregs can be created from normal CD4 T cells also if there is an immunological need. Tregs themselves are found in tumors along with anti-tumor T cells. They are supposed to safeguard against anti-self activity. In some cases, they actually inhibit the anti-tumor T cell response because of the similarities between the tumor and normal cells. Removal of these cells has been shown to increase the anti-tumor response, but a side effect of their removal is the appearance of cancer-related autoimmune responses. The Treg is generated specifically to monitor situations where there is an immune response to a known threat, such as a tumor or virus.

The other type of regulatory CD4 T cell is more universal in its function. These cells are identified by their expression of the cytotoxic T lymphocyte antigen 4 (CTLA-4), which when activated by its ligand (B-7) on macrophages induces these T regulatory cells to inhibit T cell responses of what is perceived to be anti-self (autoimmune). Removal of these cells causes widespread autoimmunity to tissues not related to the tumor. These CTLA-4 T cells seem to wander throughout the body and arrest any anti-self T cell responses. They are very important CD4 T cells because without them we would certainly die from the autoimmune reactions that they prevent. The Tregs also express CTLA-4 and some would say these two are the same, but regardless of this difference the impact of these cells on the anti-tumor cell T cell response is a major one.

Regulation of the T cell response is probably the most important aspect of the cellular immune system. Our T cell immune system must be controlled in such a way as to let it function normally, but on the other hand prevent autoimmunity from causing serious tissue damage. The effects of the loss or malfunction of these cells have shown themselves repeatedly during our immunological history. The flu pandemic in the 1920s could possibly have been caused by the inability of T regulatory cells to control the rampant destruction of normal cells by T cells that were activated by that flu virus. T regulatory cells are supposed to control this type of T cell destruction. The recent rise in diabetes and the increasing rate of arthritis may also be due to the malfunction of CD4 T regulatory cells. So, those are four CD4 T cells, and they are chameleons.

The best is saved for last. This is the memory CD4 T cell. Once an immune response is over, memory CD4 T cells are created. There's some controversy about that. Some scientists think they are created as soon as an immune response starts. Some think they are created as an aftermath of the response. We will go into detail about that later, but now it is their function that is important. The memory CD4 T cell is very much unlike the CD4 T cells that fought the battle against the original cancer or virus. It has the memory of the event embedded in its DNA. The memory CD4 T cell is different because of what is called a telomere. Telomeres are portions of the DNA that prevent the aging process. The length of the telomere equals the length of the cell's life. Nature has given the memory CD4 T cell very long telomeres. These cells can live in your circulation for an entire lifetime. Even after you die of old age the cells are still in your body.

The memory response is at least 10 times as powerful as the original one. Memory CD4 T cells are prepared to elaborate a host of lymphokines that are much more specific in terms of instructions to other cells. In reality, these are the cells Tregs were created to control. Once the memory T cells (CD4 and CD8) see the recurring infection or cancer, they destroy it and everything associated with it. Sometimes that means normal tissues as well. The Tregs come in and try to keep these cells from doing serious damage. Bluestone and Mackay (2009) elegantly review this issue of the changing CD4 T cell.

The two CD8 T cells

Now let's look at the strong-arm part of the immune system. CD8 T cells are magnificent killing machines. They have an arsenal to serve up the death of both cancer cells and virus-infected cells. This arsenal includes two very powerful tools.

The first and most lethal is the perforin attack. Perforin is an enzyme that attacks the cell wall. It forms a hole and the cell contents spill out. Actually, both CD8 and CD4 T cells can do this.

The second type of attack is called the Fas system. Killer CD8 T cells have a Fas ligand. Many cells in our bodies have the Fas receptor. When these two come together, the cell with the receptor dies. This type of cell death is not the best because what's left over is a cell with fragmented DNA. When such cells are ingested by macrophages, the antigens they present can induce a sort of immune paralysis. This effect will be covered later, but it's a real problem for the immune system. Death by the perforin system is preferred because the dead cells are processed to produce a positive T cell response to any subsequent cells of that type. Experiments are being conducted now to induce some cancer cells to express the Fas receptor, making them susceptible to this type of attack. The problem with those experiments is that they induce the formation of regulatory T cells. Once these cells are removed, protective immunity can be demonstrated.

The second of these CD8 T cells is the killer CD8 T cell. Like CD4 memory T cells, the CD8 version secretes highly active and specific lymphokines. Interleukin-2, or IL-2, released by the memory CD4 T cell activates it. There is also a need to regulate the killing activity of these cells. The CD8 memory T cell is highly lethal and can rapidly kill its target (non-self or self). T regulatory cells are often, but not always, called to the site of these CD8 memory T cell responses because they are dangerous. If they do not get there in time, severe damage to normal tissues may occur. This is a serious phenomenon, as mentioned before, with flu virus infections and a real and sometimes troublesome side effect of response to cancer. For this reason, there is an entire chapter (Chapter 7, Cancer-induced Autoimmunity) dedicated to this T cell killing spree.

One small controversy about the CD8 memory T cell is that in humans this cell has been postulated to be created from what is called a central memory T cell. The central memory cell has been theorized to have stem cell–like properties: it can change into either a CD4 or a CD8 memory T cell. This theory has yet to be proven. To prove it, T cell culture systems must be developed to produce these cells, which are only produced in the body. The CD8 memory T cell is an essential component of the T cell immune system. It is a powerful weapon used by the system to guarantee the threat will never return. Bannard and Krauman (2009) examine the dynamics of the development of CD8 memory T cells and determine that there are several distinct phenotypic characteristics that set them apart from other CD8 T cells. Also, Fearon et al. (2006) discusses the issues surrounding the development of central memory CD8 T cells.

The Natural Killer Cell

The natural killer (NK) cell is a cell of the immune system, but it is not *per se* a part of the T cell immune system. It shares quite a few commonalities with T cells: being born in the bone marrow, sharing some surface proteins, killing its target using the cytotoxic T cell (CTL) perforin system, and having a role in the immune surveillance system. The NK cell is quite different from the T cell because it does not go to the thymus after it is born in the bone marrow, has no immunological memory, and detects its target by "feeling it". Even though NK cells release lymphokines, their job is to find a suitable target cell and kill it. This is actually a very important function because dead virus-infected and cancer cells can activate specific T cell responses as a result of antigen presentation. Two separate arms of the T cell immune system handle the immune response.

The first of these arms is the "innate" T cell system and the second is the "adaptive" T cell system. The innate system is composed of T cells that have pre-determined genetic T cell receptors that recognize bacteria, viruses, and cancer cells. We are born with these. The adaptive system has T cells that learn to respond to unknown threats by producing slight mutations in those same T cell receptor genes. These mutations occur during an immune response and gradually form a receptor that recognizes the bacteria, virus, or cancer that the T cells of the innate system cannot see. This adaptive T cell immune system is the one that can cause problems because some of the mutations cause the T cell to kill normal cells. Both of these systems take their cues, so to speak, from proteins expressed by macrophages through the antigen presentation system. More about that later, but these proteins are derived from macrophages eating dead cellular debris that contains bacterial, viral, and cancer proteins. Where do these dead cancer and virus-infected cells come from? These threatening cells were killed by the NK cell (Moretta 2003).

The NK cell exists to kill from the moment it is born in the bone marrow. It may mature into an angrier killer elsewhere. After it is born in the bone marrow it finds its way, not to the thymus (checkpoint for all T cells), but to secondary lymphoid tissues such as the tonsils and lymph nodes. After a macrophage encounters cells containing non-self elements (viral and cancer proteins) they mature into dendritic cells. The dendritic cells find their way into these secondary lymphoid tissues, which are near the encounter site and release the NK cells that have been waiting there to kill something. The NK cells travel the short distance, find the non-self cells (cancer or virus-infected cell) and kill them. Then, macrophages and both arms of the T cell immune system take over. It is a very clever system because it all works quite rapidly and silently. Another interesting aspect of this system is that the NK cell hardly ever causes damage to normal tissues the way T cells can.

Natural killer cells are the immune system's bulldozers. They come in and destroy the place, and then other kinds of equipment (macrophages and T cells) arrive to clean up the mess. Natural killer cells are very important elements of the cellular immune system and should not be overlooked when designing immunotherapeutic protocols. Research is underway to determine whether these cells can be used not only in anti-leukemia bone marrow transplant therapy, but also to prevent a side effect of this type of therapy called graft vs. host disease (GVHD). This disease occurs when the T cells of the donor attack normal tissues of the leukemia patient who is the recipient. It can cause significant sickness and, in many cases, death. Because NK cells secrete suppressive lymphokines, they can stop the action of the T cells responsible for GVHD. It was found that adding donor NK cells (that were stimulated) to the donated bone marrow resulted in the reduction of GVHD. So in the end, as far as immunotherapy is concerned, the NK cell may very well be the shining star of the immune system. Berzofsky and Terabe (2009) discuss the role the NK cell plays in the generation of an anti-cancer T cell response, and all of Chapter 12 in this present work is devoted to these very important cells.

Cells of the Monocyte-Macrophage Series

The cells of the monocyte-macrophage series are a series of related cells and, like the T cells, subdivided into distinct cells with specific functions. If CD4 T cells were chameleons, these cells would be the masters of change. In many cases the change is permanent. Cells of the monocyte-macrophage series are born in the bone marrow and go directly into the circulation, where they are in constant motion. However, when this motion stops and they settle in one of the many tissue systems in the body, then the real changes occur. In some instances they incorporate the characteristics of the tissue they settle into, until it is hard to distinguish them from the resident tissue. This is called differentiation, which literally means change into something else.

Differentiation will be covered in detail as we go along. One interesting thing is the physical change these cells go through when they stop moving. A T cell is 96% nucleus (DNA) and 4% cytoplasm (protein), but a still monocyte-macrophage is about 30% nucleus and 70% cytoplasm. The nearest thing you can compare them to is an amoeba. They are highly mobile because they can squeeze between cells in solid tissue. They also can engulf (phagocytize) cellular debris and particulate matter the same way an amoeba does. Through this critically important function, these cells make proteins available to the immune system.

The cells of the monocyte-macrophage series control a wide range of functions throughout the immune system. They can positively or negatively affect an immune response through a variety of methods. To do this, they can use the secretion of short-range small molecules like lymphokines. In addition, they can directly contact another cell and signal it to do something using the hundreds of proteins it expresses on its outer cell membrane. Their most magnificent function is their ability to present foreign protein antigens to T cells. This function is called antigen presentation. It is a process that involves hundreds of biochemical steps. The end result is that what is now called macrophage tells the T cells what cellular threats to attack based on the proteins it has extracted, processed, and presented. This is a central function of the immune system, and without it there would be no immune system.

The monocyte

The monocyte is actually the first of its kind to enter the body from the bone marrow. Monocytes circulate throughout the body. They can freely enter and exit the brain. This is no small feat because there is a blood-brain barrier, a gate to the brain and all the cells within it, that excludes most cells and chemicals. This barrier presents problems when administering chemotherapy for cancer and HIV. Some T cells are selected to enter the brain under special circumstances, but the monocyte-macrophage has no problem in this regard. Once the monocyte stops in the brain or any other tissue system, it immediately begins the process of differentiation. This process of differentiation not only turns the monocyte into a macrophage, it also does something else quite extraordinary.

Each tissue system in our bodies is made up of a certain type of cell. For instance, the nervous system has neurons, the liver has hepatocytes, and the skin has epithelial cells. During this process, the monocyte takes on the characteristics of the cells in that tissue. If they stop in the nervous system they are called microglia cells. In the liver they are Kupffer cells and in the skin they are called Langerhans cells. Early histologists were unaware that many of the cells they were looking at were not derived from the tissue they were studying, but in fact were macrophages that took on the appearance of the resident cell population. However, this change the macrophage goes through means that it may never leave the tissue it settled in. Other macrophages may come in and go out of the same tissue, but the resident macrophages must stay put because they serve a special purpose in that tissue. So the monocyte is the starting point for the continual and specific changes these cells go through to support the immune system as well as the entire biological system.

The macrophage

If the monocyte is the starting point then the macrophage is the intersection. As mentioned earlier, macrophages differentiate once they reside in the tissues. Depending on the stimulus it receives, the macrophage can turn into an aggressive immune cell capable of dictating the very nature of an immune response. These aggressive macrophages are called dendritic cells. It's important to realize that the macrophage itself is a powerhouse and, in some respects, resembles a factory. It has a huge amount of cytoplasm for a reason. One of its functions is to process antigens for the T cell immune system. Antigen processing is just a part of the antigen presentation function.

To briefly look at how macrophages present antigens, let's define what an antigen is. An antigen is usually a small segment (peptide) of a protein that is expressed by a bacteria, virus, or tumor that the body's T cells recognize as foreign. Antigen presentation will be covered in detail in Chapter 2. However, some physical aspects of this process that are unique to the macrophage warrant comment here. Let's say a macrophage stumbles on some cellular debris that was the result of the NK cell blowing to bits a virus-infected or cancer cell. Like an amoeba, it surrounds the debris with its cell membrane and forms a vesicle inside the cell. A vesicle is like a bubble floating inside the cell. These vesicles fuse with other vesicles containing an enzyme (proteolytic enzyme) that breaks down proteins into subunits called peptides. The peptides are usually 8 to 10 amino acids long.

T cells have a genetic receptor for foreign peptides, but the peptides must be associated with a self-protein. This is a complex to be expressed on the surface of a macrophage for CD4 T cell activation. On the other hand, this complex must be expressed on the surface of a virus-infected or cancer cell in order for CD8 T cells to kill that cell. Inside the macrophage, there are other vesicles that contain these self-proteins (HLA-D) and they, in turn, fuse with the ones containing the non-self peptides. The peptide and self-protein associate and then this new vesicle fuses with the outer cell membrane, exposing the complex. CD4 T cells with the appropriate genetic or adaptive receptor see this and are activated, which is the start of the T cell immune response.

Aside from their immunological function, macrophages also perform a housekeeping function that is important to the entire biological system. They actively clean up debris wherever it is found regardless of what it is. Macrophages do not usually discriminate about what they ingest, they just eat it. Sometimes they eat so much debris that they gather together in a mass called a granuloma. This condition is often associated with infections of bacterial or fungal origin. It is a special type of inflammatory reaction in which the macrophage is trying to isolate the infection from the body by forming a solid mass. The macrophages in this mass often take on the appearance of epithelial cells of the skin and in some cases are mistaken for cancerous masses. However, these granulomas are quite harmless, although they can be a problem because of their size. The macrophage just by itself is quite an amazing immune cell. However, the next step in its differentiated life is one that commands respect because of its power over the immune system. That cell is called the dendritic cell.

The Dendritic Cell

Dendritic cells (DCs) can be considered "aggressive directors" of the T cell immune system. They are called "aggressive" because of how they are formed and what they do once they are created. Dendritic cells can be created directly from monocytes or from macrophages. They are end-stage specialized cells of the monocyte-macrophage lineage. Ninety-seven percent of their function is related to the immune system and activities related to T cell immune responses. Dendritic cells are aggressive compared to their macrophage counterparts. Their physical appearance is a testament to that. Once the DC is created, it barely resembles the amoeba-like macrophage. For instance, the nucleus is much larger: it occupies about 70% of the DC instead of 30% in the macrophage. This extra nuclear material is needed because of the amount of instructional immune-directing proteins it must make. It still has the ability to ingest cell debris, but it is a little less mobile than its parent, the macrophage. Because of what the DC has to do, it does not have to squeeze through solid tissue, but goes to sites where it is needed through capillaries or the lymphatic system. The dendrite takes on a spike-like appearance and is able to send out slender tendrils, much like nerve cells. It uses these tendrils (hundreds per cell) to communicate with other cells and particularly T cells. This cell can do things that no other immune cell can, despite its aggressive mission-like role.

The DC is actually created by T cells responding to immune threats, such as certain types of bacteria, viruses, and cancer cells. When activated by one of these threats, the T cell (CD4 T cell) secretes a special combination of short-range proteins called lymphokines. One lymphokine is called granulocyte-monocyte colony stimulating factor and the other is called IL-4. When these two proteins are detected by either monocytes or uncommitted macrophages they gather at the site and are transformed into DCs. Once the DCs are formed, they begin to communicate with T cells at the site after sampling the proteins associated with the immune response that is occurring. The T cells generally seek out DCs for reasons that will become apparent later on. However, some of the DCs leave the site after they take a protein sample. They naturally travel through the lymphatic system, but sometimes they take short cuts via the bloodstream. Invariably, they wind up at some local or distant lymph node. Here, as mentioned before, while in the lymph nodes they release NK cells that have been waiting there to kill something. They also make contact with some of the millions of T cells, which reside in the lymph nodes and travel through the lymphatic system.

What can DCs do as far as the T cell immune system is concerned? First of all, the transformation from monocyte-macrophage to DC causes the machinery of the antigen presentation system to work faster and more efficiently. Unlike its parent, every single DC has the ability to present antigens, and it does it better and faster by a factor of 20. If a DC is created in the lab, using the lymphokines mentioned above, it can be used to induce killer CD8 T cells. Once antigen (viral or cancer) is added with IL-2, the DCs quickly mature naïve CD8 T cells into killer cells in hours. The DCs can in a few minutes turn cancer- or virus-infected cell debris into immune-recognized peptide/self-protein complexes. The macrophage can do this, but in some cases it takes five times as long as the dendrite does. That's how efficient the DC is at presenting antigens. Another equally amazing function is its ability to guide the immune response down one path or another.

As mentioned before, there are two types of CD4 helper T cells. Type 1, which controls CD8 killer T cell activation and type 2, which controls antibody responses while suppressing the CD8 killer activation phase. The DC plays a major role in the selection of the type of CD4 T cell that dominates the response. Dendritic cells have shown themselves to be aggressive, determined, and very specific in what they do. They play a central role in the T cell immune system's ability to fight cancer. Dendritic cells can send multiple signals to a single T cell. The signals tell the T cell not only what to do about the cancer, but also how to affect the rest of the T cell responses that occur afterwards. For instance, DCs can shut off a T cell immune response. They tell the T cell to kill a certain type of cell (virus-infected or cancer) and then shut down any subsequent T cell activation events related to that entity. If the T cell immune response is not shut down after the threat is gone, the T cells left over will be continuously stimulated by cellular debris that was a result of the attack. This situation would invariably lead to an autoimmune reaction, which could seriously damage normal tissue.

As mentioned before, the adaptive T cell immune system works through slight mutations in the T cells receptor. This can occur in these "after the response" T cells and it can generate a receptor that is anti-self. Therefore, the DC sometimes instructs a T cell to be both a killer T cell and a T regulatory cell at the same time. This function can also work against an anti-cancer response because once a certain population of cancer cells is destroyed there is often another being created at the same time because of the stem cell nature of cancer. When this new population of cancer cells is created, the T cell response, which could potentially attack the cancer, has been suppressed by the instructions given by DCs through their ability to create T regulatory cells.

Scientists and clinicians are exploiting the acute ability of DCs to present cancer antigens to T cells. In addition, these cells can easily be created in the lab. This makes them perfect candidates to be used in the development of anti-cancer immunotherapeutic protocols. Clinical trials are underway using DCs as "cellular vaccines". In this method, a patient's cancer is removed and turned into cellular debris in the lab. At the same time, the patient's monocytes are converted into DCs. The patient's DCs are incubated with this cancer cell debris (protein antigen) and they begin the process of antigen processing and presentation. Next, the antigen-bearing DCs are returned to the patient's blood. The patient is then given IL-2 intravenously. That matures the activated CD8 T cells to become anti-cancer killer cells.

What has just been described is the core of many variations of this type of immune-based therapy. The results have been encouraging in end-stage patients. However, the problem of regulatory T cell control of the anti-cancer T cell response still remains. A complete response would remove the cancer and leave memory T cells behind to guard against its recurrence. This will not happen until scientists can block the T cell regulatory signal the DC provides to the anti-cancer killer T cell that turns it into a Treg. Clinical trials of this technique are now being carried out on patients who are at the initial stages of the disease and it is hoped that the problem will be solved in time to stop the spread (metastasis) of cancers like melanoma and brain tumors like gliomas.

The DC, unlike many of the cells of the immune system, is a workhorse. The results are encouraging, but not yet at the stage where they are curative. In the future of immune-based cancer therapy, these cells will be used in combination with other cells of the immune system to generate a comprehensive total and final immune response that would result in complete removal of all cancer cells, including the cancer variants that are created by cancer stem cells. The stem cell nature of cancer is one of the biggest challenges to the T cell immune system's ability to fight cancer. However, the cancer cell itself has many other ways to not only evade the T cell response, but also confuse it. Stout et al. (2009) reviews this issue of the ever-changing monocyte-macrophage and their participation in the anti-cancer T cell response.

The Cancer Cell

Cancer cells themselves play a pivotal role in the T cell immune response, which is supposed to remove the cells. Many cancer cells begin life as slightly altered normal cells. These abnormal benign cells can actually function normally for a long time. At first they do not proliferate (divide) uncontrollably, produce mutant proteins, or disrupt the microenvironment they are in. What usually separates them from their normal counterparts is the complexity of their nucleus. Many times, nuclear DNA mutations that ultimately lead to cancer result in the production of extra DNA. Therefore, the benign precancerous cell is what is called a multiploid or aplastic cell. This happens because the breaks in the DNA are repaired incorrectly, which leaves that segment of DNA unable to be duplicated; it is just left over or, worse, duplicated.

As time goes on two major events occur. The first is the joining of DNA (through the repair mechanism) of two separate protein-coding regions. This results in the formation of hybrid proteins previously unknown to the T cell immune system. The proteins are either expressed on the precancerous cell surface or secreted into the body. Either way, the immune system is alerted by this event. The NK cells also can feel these irregular proteins on the cell surface. The other and probably the most deadly event is the disruption (by mutation) of genes that regulate cell growth. Once these genes are altered to the point where they can no longer do their job, the cell begins to grow in an unregulated fashion. A cancer cell is created.

Now the real problems begin for the T cell immune system. However, sometimes (more often than we think) it seems the cancer cell wants the T cell immune system to see it. As mentioned before, CD4 T cells can only respond to non-self peptide antigens unless they are associated with a self-protein on the antigen-presenting cell surface. Twenty years ago, I and a group of scientists and clinicians (led by Dr. DuPont Guerry) at the University of Pennsylvania discovered something quite extraordinary about cancer cells (melanoma) in the early stages of development. We found that benign and early (primary) cancer cells could process and present antigens exactly like normal macrophages. This function, performed by primary melanoma cells, had identical attributes possessed by cells of the monocyte-macrophage series and, in fact, the cells could present traditional peptides like the ones derived from tetanus. The exception here was that these cancer cells were presenting melanoma cancer antigens recognized by the T cell immune system. That is why patients with this type of melanoma had a 95% survival rate in contrast to patients with metastatic melanoma (which was unable to present antigens), who had a 5% survival rate. Alexander and co-workers (1989) discuss this phenomenon as it relates to the prognosis of patients with these two distinct forms of melanoma. This type of cancer develops from tissues of the nervous system and there is an entire chapter devoted to the specific problems the T cell immune system faces when responding to these types of cancers.

Eventually, we theorized that this was a double-edged sword. On the one hand, it had to lead to the destruction and elimination of many newly formed cancers. On the other hand, in some cases, it was a way to invite the T cell immune system in only to confuse it and suppress any subsequent attempts to remove it. This confusion and suppression of the T cell immune system is the hallmark of the most deadly cancers. It is a very clever method that encompasses a variety of mechanisms used by the cancer cell to dominate not only its surrounding area (microenvironment) but also the entire body. This occurs well before the presence of a cancer can be detected. As a result, the T cell immune system is already paralyzed, suppressed, or just plain confused by the time the cancer is diagnosed. The initial decisions made during the first few encounters of the T cell immune system with the cancer can dictate the outcome or ultimate prognosis (fate of the patient). Once the cancer cell expresses antigens recognized by the T cell immune system, the invitation is put out, but the response to the cancer may not be a good one.

This invitation of the T cell immune system to the cancer site comes with a price. The cancer cell can alter the T cell immune response so that the T cell is unable to fight not only the present cancer, but also its subsequent incarnations. It has been known for some time now that cancer cells behave almost exactly like stem cells. Stem cells are primordial cells that can turn into any cell in our body depending upon the stimulus acting on them. Short-range proteins called lymphokines and others called chemokines program stem cells. Once that happens, the stem cells differentiate into a particular cell type and stop dividing. Unlike the normal stem cell, the cancer stem cell does not stop dividing and reproducing itself.

The cancer stem cell again, unlike its normal counterpart, is not influenced by short-range proteins, but by the constant appearance of mutations, which create new cancer cells in the form of sub-populations. These cells can simultaneously produce several distinct sub-populations of cancer cells. Each of these has different protein configurations and growth characteristics. Therefore, if the T cell immune system gets confused or suppressed by the early cancer, it is definitely at a distinct disadvantage when these new cancer cells are created. How early cancer cells do this is quite interesting and happens at multiple levels of the T cell immune system.

The cancer cell has a wide variety of specific methods that confuse and even shut down the T cell immune system. Cancer cells are masters at misdirecting T cells. Before this cell became cancerous it was quite normal and ignored by the immune system. Cancer cells can create hybrid proteins (tumor antigens) that, when expressed on their surface, not only alert the T cell immune system, but also are targets for cytotoxic CD8 T cells. Unfortunately, in the early stages of cancerous development, normal proteins are expressed side by side with these abnormal tumor antigens. This is a bad situation for the T cell immune system. There are two very confusing states that can occur: The first has to do with a failsafe mechanism the T cell immune system has. Some normal cells in our bodies display very slight modifications in protein expression that are generally associated with aging and normal differentiation. The T cells see the modified cell, but instead of attacking it they "pause" and allow the cell to stabilize and continue its existence. This mechanism also comes into play when fetal proteins are detected during pregnancy. They do not attack the tissue and instead they wait. Sometimes the dual expression of normal and tumor antigens activates this failsafe mechanism. In the meantime, when the T cells are in the "pause" mode, the cancer cells quietly increase their numbers. This makes the situation much worse for the T cells when they finally wake up and start to fight a cancer that has all of a sudden outnumbered them.

The second situation has to do with the accidental induction of autoimmunity. If the T cells respond not only to the tumor antigens, but also to the normal antigens, a separate T cell immune response may be induced. This T cell response is not directed at the cancer, but at the T cells inadvertently attacking normal tissues. The T cells from the initial response think they are attacking the cancer, but in fact they are damaging the normal cells the cancer developed from. This happens in numerous types of cancers and, in fact, may happen in all of them. The cell performing this attack on other T cells is called the T regulatory cell or Treg. The Treg has to go into the cancer mass and sort out this situation of two killer T cells with different targets. The idea is that the Treg has to stop the anti-self T cells while leaving the anti-cancer T cells to do their job, which is to kill the cancer cells. That's how it's supposed to work. Unfortunately, it doesn't always. The Treg has difficulty distinguishing between the anti-self and the anti-cancer T cells.

Often, both cells are inhibited from responding, but eventually the anti-cancer T cells break free of this control. Before that happens, the cancer continues to grow. Often, once the Tregs have sorted things out, the cancer cells vastly outnumber the available anti-cancer T cells. One good thing about this is that during that time DCs have arrived on the scene and they quickly begin to convert the resident Tregs into anti-cancer T cells. This can be a cyclical phenomenon, which begins as the T cell response again spawns anti-self killer T cells. So, as you can see, the cancer cell can cause havoc with the T cell immune system. There are several other ways the cancer cell can subvert the T cell response by the very expression of these hybrid proteins.

The T cell responds to a small protein subunit called a peptide. This peptide is presented on the surface of a macrophage or target (cancer or virus-infected) cell in association with a self-protein. Sometimes, when the cancer makes these hybrid proteins, the peptides produced by the antigen-processing step are actually either half a peptide or half T cell-stimulating peptide and half something else. In other words, they are altered peptides. When T cells respond to these altered peptides, they induce a partial signal to the T cell. This can happen to any responding or killer T cell. The partial signal initiates a negative response that tells the T cell to secrete immune-suppressing lymphokines. This is called an anergic T cell, a suppressor T cell. It is separate from the Treg, but the results of the suppressor T cell are the same. This T cell response to tumor peptides and their use as anti-cancer vaccines is discussed by Jordan and co-workers (2010). These lymphokines not only suppress other T cell responses, but also cause macrophages or DCs to do something really bad.

Under these conditions, macrophages will secrete a compound (IDO) that interferes with the T cell's ability to process tryptophan. T cells need this amino acid to function, and without it they are paralyzed. When macrophages secrete this compound, T cells in the area are shut down. Even if the T cells escape the cancer mass where this is occurring, it takes some time for them to recover. Von Bergwelt-Bailton and colleagues (2006) discovered that tumor-secreted prostaglandin E2 (PEG-2) induced IDO release from DC. So, in a sense, this type of T cell suppression is systematically bad and may affect other immune responses not related to the cancer. That is why many cancer patients, even without receiving chemotherapy, are prone to infections. It is a generalized suppression of the immune system induced by the cancer cells with the express purpose of evading the T cell immune system. This phenomenon can occur very early in the development of the cancer, well before it is detected medically.

These are not the only ways a cancer could evade the T cell response to it. As we have said, cancer cells possess an arsenal of weapons against the T cell almost as effective as the ones the T cell immune system has against it. The defensive systems of cancer cells are complex and sometimes quite subtle. Sometimes cancer cells deploy some of these defenses as soon as they are formed. This puts the T cell immune system at a distinct disadvantage early on. In order to design effective immunotherapeutic treatments for many forms of cancer, these actions by the cancer cell must be taken into account. It would also help to develop a new generation of screening methods that would detect the byproducts of the cancer's defensive systems. This would allow for very early detection of a cancerous event in the body.

From these various methods used by the cancer to defeat the T cell immune system, it may appear that there is no immune response at all. However, more often than not there is a T cell response, but by the time the cancer is detected medically, the T cell response is all but extinguished. This is where immunological intervention is needed. Consequently, the cancer cell itself is being used to develop new immunotherapeutic treatments. The "cellular vaccines" treatment mentioned earlier is just one of the many ways the cancer cell is used to stimulate the T cell immune system. In the future, when immunotherapy becomes standard treatment for cancer, extraction and use of a patient's cancer cell will be a component of an integrated immunotherapeutic treatment plan. In the end, however, manipulation of these cells of the immune system will be the only way to remove the cancer and keep it out. All the elements of an integrated immunotherapeutic approach to the treatment of cancer are being tested in patients in clinical trials. Only when the various parts are put together will the real balance of power shift toward the T cell immune system. Prendergast and Jaffee (2007) review the basics of this new anti-cancer treatment.

Conclusion

What you have just read is an overview of the role the cells of the immune system play in the ability of the T cell immune system to remove cancerous cells from our bodies. The information about these cells is presented in a segmented manner, but they work together and have functional relationships with each other. The cells work in concert, one type of cell picking up the battle after another type has finished. The amazing thing about all of this is that it is practically automatic and happens many times without our knowing it. Once the cancer cell is detected, this initiates a cascade of events that is supposed to lead to the destruction of the cancer. More importantly, however, if the response is a success, a living memory is left behind to safeguard against the recurrence of that type of cancer. The cancer itself has formidable defenses and has the ability to counter potential attacks against it by the T cell immune system. Cancer can disrupt this symphony of the immune system as it goes about trying to destroy it. However, each one of these cell types has the potential to be modified (by immunotherapy) in such a way as to circumvent these attempts made by the cancer to take over the body. There is an order and connection between the cells of the immune system that makes all this possible.

For example, NK cells and cells of the monocyte-macrophage series have a special relationship with each other during the development of a new cancer. This relationship directly affects the way the T cell immune system attacks the cancer and is a critical step toward its removal. Often, circulating NK cells see newly formed cancerous cells and kill many of them. The results or cellular debris left over are processed into T cell–recognizable peptides and presented to the T cell immune system by macrophages. The response by the T cell immune system to these peptides triggers the formation of DCs. In turn, the DCs release even more NK cells from the lymphatic system, increasing cancer cell death, which results in accelerated activation of the T cell immune system. This type of reaction can lead to a full-blown attack by the T cells on the cancer cell population. If that is done soon enough, the cancer stem cells have no chance to spawn an "immunologically invisible" type of cancer cell.

Looking at just this one pathway leading to recruitment of an anti-cancer T cell immune response, it is apparent where immunological intervention could be applied to tilt the balance of power toward the T cell immune system and the inevitable destruction of the cancer. For instance, by detecting the cancer at a very early stage (say 1,000 cells) with methods not yet developed, it may be possible to do two separate things that would accelerate the anti-cancer T cell response and quickly remove these cancer cells. The first would be to stimulate NK cell production, which would make more cancer peptides available (via antigen presentation) for the T cell response, or to remove NK cells from the patient and reproduce them in the lab, then return the increased number of NK cells to the patient with the same result. The second, either alone or with the above procedure, would be to treat the patient using the granulocyte-monocyte colony stimulating factor lymphokine to increase the formation of DCs, which translates to increased antigen presentation to the T cell immune system.

The balance of power between the T cell immune system and cancer is at times quite precarious and tenuous. It can be tilted by both internal and external influences. As far as internal influences are concerned, anything that diverts the T cell immune system from its primary goal of eliminating the cancer is not good. The T cell immune system is like an army and has limited resources. Sometimes these resources are diverted by opportunistic pathogens. An example is the JC virus (JCV), which can be invisible to the immune system. This is a latent virus carried by over 90% of the adult population. Normal, healthy individuals are not affected by a viral outbreak because the immune system keeps the virus latent. However, if a person has AIDS, cancer, or an organ transplant, which suppresses the immune system, a JCV outbreak occurs. JCV attacks neural tissues and causes a fatal neurological disease. It also can cause cancers in nerve cells. Unfortunately, most of a person's T cells are fighting the cancer. In a sense, the patient's T cell immune system has been compromised by divergence of T cells to fight the cancer. The T cells not fighting the cancer are called upon to attack this virus, but there are not enough of them. Recruitment of additional T cells to fight the cancer becomes a problem, but the other problem is that JCV-infected cells can suppress T cell function. Gallia and co-workers (1997) were among the first groups that focused on the immune-inhibiting ability of this virus. Here, because T cells are being inhibited, the balance of power is shifted toward the cancer.

An external factor that can influence this precarious balance of power is chemotherapeutic treatment of the cancer itself. Chemotherapy (covered in detail later in the book) is toxic not only to the cancer, but also to T cells. Many cancer patients being treated with traditional chemotherapy develop leucopenia, which means diminishing numbers of white blood cells (including T cells). Often, white blood cell stimulators are used to increase the number of immune cells to prevent fatal infections. With this type of chemotherapy, the T cell immune system can be diminished to the point where it no longer functions properly. Fortunately, the newest generation of chemotherapeutic agents target cellular pathways only the cancer cell possesses and are much less toxic to T cells. Scientists are now trying to design these drugs so that they produce a "good" type of cancer cell death. The T cell immune system can use the peptides from these dead cancer cells as a vaccine of sorts. If such agents are developed properly, the T cell immune system will be able to kill the remaining cancer cells and leave memory T cells behind. Using chemotherapeutic agents like this, rather than traditional chemical agents, would shift the balance of power toward the T cell immune system.

So, which cells control the anti-cancer T cell immune response? Let's start with what I consider the supporting cast. The NK cells are definitely important because they make available cancer cell peptides through their ability to instantly create cancer cell death. This death leads to the generation of cancer cell debris used by the macrophage to generate the cancer peptides recognized by the T cell immune system. Important as this activity is, it is somewhat redundant because macrophages can cause cancer cell death as well. The advantage of the NK cell system is that it is a surveillance mechanism, where the NK cell stumbles upon a cancer cell and begins the cancer recognition-response cascade. The NK cell, like many of the others, gets high marks for its unique role in the T cell response to cancer and because it can be used for immunotherapeutic treatment of cancer.

The next most important and absolutely necessary cell type are the cells of the monocyte-macrophage series. These cells play a critical role in that they provide the T cell immune system two things: The first is the stimulating cancer peptide, which in a sense is a landmark that the T cells use to direct their attack. The second is direction to the T cells in the form of immuno-modulating signals delivered by cells such as the DCs. As a group, the cells of the monocyte-macrophage series are a powerful force. They can kill cancer cells as well as prepare the T cell immune system to attack it. It is hard to argue that without them there would be no immune response. Their non-immune functions alone elevate them to one of the most important cells in the body. Again, the DCs are so important that they have been used extensively in immunotherapeutic clinical trials in humans. However, because of the complex relationship these cells (particularly DCs) have with the T cell immune system, their effectiveness has not lived up to expectations. Once we solve the problems related to DC induction of T regulatory cells (which impede anti-cancer T cell responses), DCs will be used to cure the cancer.

Now we are left with the final group of cells, the T cells of the cellular immune system. These cells control the T cell response to cancer. T cells, as a group, are the final solution to the cancer problem for two reasons. First, once they have identified the cancer through recognition of its peptides, the T cells can go about eliminating all of them via the killer CD8 T cell. The second reason is this generation of the living memory of the cancer cell itself. Alone, these two factors are the immune system's most powerful weapons against cancer. The question of how T cells control their own destiny in fighting cancer is a bit more complex.

The T cell response is actually broken down into two parts. The first is the initial CD4 T cell response. The second is the phenomenon of T cell regulatory cells and the way they police the T cell immune system. The initial CD4 T cell response is quite critical, because it basically sets the tone or type of T cell response that follows that first encounter between the naïve CD4 T cell and the antigen-presenting cell, be it a macrophage or cancer cell. Several events that can occur at this point cause that CD4 T cell to either marshal the full muster of the T cell immune system toward a positive response, slow it down, or prevent any subsequent T cell immune response from occurring. Conditions such as partial peptide presentation, altered presentation, and the fetal failsafe mechanism induce a negative T cell immune response to the cancer from the very beginning. This is in contrast to a positive response that recruits not only killer T cells, but also a host of other cells, such as DCs and NK cells, to join the fight against a growing cancer. This form of control is about pre-determination of the anti-cancer T cell response, which in most cases cannot be changed once it starts. Manipulation of these events must somehow be factored into the design of immunotherapeutic treatments. Without control of this aspect of the T cell response, the race to eliminate the cancer is lost almost before it starts.

The second and probably the most important aspect of how T cells control the T cell response to cancer is T cell regulation. This happens well after the initial response, but it can be just as debilitating to the T cell immune system as that initial negative signal delivered by the naïve CD4 T cells. CD4 T cells called Tregs are specialized T cells identified by the expression of a gene called FOX3 (Elgert 2009). This is just one of the two CD4 T regulatory cells that basically police the T cell immune system. Again, Tregs are mentioned because they often appear or are created at the sites of cancer. What they do is monitor the activities of the anti-cancer T cells and make sure that as these T cells are created none of them begin to attack normal tissues. As anti-cancer T cells develop, some often attack normal tissues that the cancer developed from. When this happens, the

Treg comes in and shuts down that response. Unfortunately, sometimes this inhibition of self-reactive T cells affects the anti-cancer T cell response. It can slow down the anti-cancer T cell response so much that the cancer cells grow unabated and can very quickly threaten the life of the patient. This situation is even more complicated because of the very nature of the Treg cell.

The CD4 regulatory T cell, or Treg, can exist in a "fluid state". Tregs appear in situations where autoimmunity is likely to occur. Their job is to stop these anti-self T cells. It turns out that inflammation is a precursor to autoimmune reactions. For example, an arthritic person's joints become inflamed and normal tissues (nerve and cartilage) are damaged. This causes serious pain and swelling. Inflammatory conditions can turn normal or even immune CD4 T cells into Tregs. This happens because when macrophages detect an irritant, such as bacteria or bodily fluids like collagen, they release inflammation-producing proteins, such as chemokines and specialized lymphokines. These proteins act on the CD4 T cell and convert it into a Treg because that type of T cell is needed under these conditions. Radhakrishnan and colleagues (2008a) discuss this fluidity of Tregs and the implications related to autoimmunity.

What does this have to do with cancer? Some cancers, such as brain tumors (gliomas), begin to grow unnoticed by the T cell immune system initially and begin to crowd out normal cells. These normal cells release irritants and the macrophages come in. Inflammatory conditions are created and CD4 T cells, even if they are anti-cancer T cells, begin the process of being converted into Tregs. This happens simultaneously or separately. In the end, there is massive confusion on where the killer T cells are. They can kill the tumor on the one hand, but on the other hand they inhibit the development of new anti-cancer T cells such as CD8 killer cells.

What is more confusing is that this phenomenon can happen in reverse and it can result in the destruction of normal tissue. In that case, as the cancer forms, it is immediately recognized by the immune system. CD4 T cells induce a productive T cell response that awakens and changes immune cells in ways that make the job of killing the cancer easy. Tregs find their way into the cancer site because that is what they are supposed to do. The CD4 T cells in this situation secrete a combination of lymphokines that turn the macrophages at the site into DCs. As the cancer continues to grow and the T cell immune response continues to increase, there is a constant need for the development of new anti-cancer T cells. This is where the DCs come in. Seeing the shortfall of anti-cancer T cells, the DC does something amazing. It begins to convert any T cell into an anti-cancer T cell. This is a very special function of DCs and is thought to be used only in emergencies, where the cancer begins to outnumber its opponent, the anti-cancer T cell. Apparently, two things have to happen: the DCs must secrete a special lymphokine and they must make direct contact with the T cell that is to be converted.

Fortunately or unfortunately, the Treg becomes one of the T cells destined to become an anti-cancer T cell. Again, and for a while at least, there is dual expression of the two functions (T cell regulation and anti-cancer responder) until full conversion is achieved. This loss of the T cell regulatory function during the heightened anti-cancer T cell response can cause serious problems for the patient. During the periods of rapid cell division during a T cell response, mutations occur in the genes of the anti-cancer peptide receptor the T cells use to see the cancer. Some of those mutations result in the creation of anti-self T cells that attack tissues usually, but not always, related to the tumor. Now, there are very few Tregs available to control these anti-self T cells, and usually patients suffer debilitating side effects. As a matter of fact, the cancer sometimes is discovered because there is an autoimmune reaction to a related tissue system.

So you may consider this a "yin and yang" situation, where functional characteristics (T cell regulation) appear and disappear, depending on conditions at the cancer site. This type of control of the anti-cancer T cell response is unique to the T cell population. None of the other immune cell types have this ability to control members of their own population. Because of situations like this one and the one concerning the initial CD4 T cell response, it is clear that T cells control their own fate as far as the anti-cancer T cell immune response is concerned. The T cell is by far one of the most complex and adaptable cells produced by our bodies. In some respects, only the cancer cell can come close to it. Our lives literally depend on the fact that the T cell does have the ability to out-maneuver and destroy the cancer. Through the use of immunotherapeutic treatments, the hope is that this population of immune cells can permanently rid us of cancer.

Chapter 2

Antigen Presentation

Antigen presentation is a complex biological process that is at the root of the basic inner workings of the immune system. This process can determine if a disease such as cancer or a viral infection can survive and flourish or be removed from the person's body, never to return. The antigen presentation system is an essential component of our immune system. Without it we would succumb to the most basic infections, from bacteria to fungi. Pathogens that normally have no effect on us would be deadly. Cells of the immune system are not the only ones capable of antigen presentation. Indeed, this function is the basis of our ability to recognize the difference between self and non-self. This is the fundamental function of our entire immune system. Autoimmune disorders, which are numerous, make use of this pathway to display self-antigens that are inappropriately recognized by the immune system. We must study antigen presentation thoroughly if we want to understand how the immune system works and how it can be manipulated to halt anti-self reactions, destroy tumor cells, or eliminate viral infections. The way the immune system works and the factors that control its function are pivotal to our ability to understand immunity.

In this chapter we first take an overview of antigen presentation. Then we examine the cellular proteins that facilitate antigen presentation. After this, we describe the main cellular recipient of the process, the T cell. We highlight the role antigen presentation plays in autoimmunity and suppression of immune responses. We explain in detail how antigen presentation functions at the protein level and how the parts of the system are assembled. We then use this information to describe how tumor cells and viruses subvert the T cell immune system. Finally, we discuss new advances in antigen presentation research as they relate to the future, where immunotherapeutic treatments will be used to fight tumors and viruses.

What Is Antigen Presentation?

In basic immunological terms, antigen presentation is the ability of several populations of antigen-presenting cells (APCs) to signal or direct other populations of cells (T cells, B cells, and NK cells) to seek out or respond to small chains of amino acids (peptides) that are parts of larger proteins. These proteins can be a part of a vast number of entities, such as bacteria, fungi, toxins, viruses, other cells (cancer), air-borne debris (allergens), and a host of other protein sources. There are two basic forms of antigen presentation. Each form is associated with two cellular protein families that identify our cells as self. These are termed major histocompatibility antigens (MHC).

The Major Histocompatibility Antigens

MHC class I

The first type of histocompatibility protein is the MHC class I protein antigen termed human leukocyte antigens or HLA in humans (H-2 in mice). This family of proteins comprises an alpha and a beta subunit. Genetically, these proteins are inherited and expressed as a mixture of parental proteins. Each individual has a unique combination of these proteins that distinguishes that individual from others. The proteins are expressed on virtually every cell in the body. The antigens, expressed by epithelial cells in the thymus, are essential for the formation of our immune system. The reason for this is that the major cell of the immune system, the T cell, is instructed by the thymus not to respond to self. If T cells respond to self-antigens, which are expressed by epithelial cells in the thymus, through the T cell receptor (TCR), they are eliminated. Doherty and co-workers (2008) discuss MHC class I and its relation to T cell diversity.

The response to non-self antigens is genetically embedded in the genes that make up the receptor itself. When these receptors (expressed by CD8 T cells in the HLA system) are engaged via the MHC class I/peptide on the surface of the APC, molecular changes begin to occur inside the T cells that allow them to release specialized proteins (lymphokines), or rapidly divide, or mature into a killer (cytotoxic) or memory T cell. The peptide antigen that is presented in this case is processed through an endogenous (internal) pathway and is associated with the HLA class I molecule during its synthesis within the APC. In contrast, there is an exogenous (external) pathway of antigen presentation associated with our next class of MHC antigens (class II).

MHC class II

The second form of antigen proteins is the MHC class II protein antigen, called HLA-D in humans (H-2D in mice). The proteins are similar to HLA in that they are inherited and structurally have alpha and beta protein components. However, there is a difference in expression, in that HLA-D antigens are expressed by a subset of cells in the body. The cells that express the antigens are B cells (antibody-producing cells), monocytes (circulating lymphocytes), macrophages (mature monocytes), dendritic cells (very mature monocytes), Langerhans cells (tissue-specific), and microglia cells (nervous system) and, in some cases, activated T cells. In addition, there are cells in the bone marrow that naturally express class II antigens. Therefore, class II expression is not as widespread as class I antigen expression. The TCR of the CD-4 T cell recognizes stimulatory peptides presented by cells (APCs) that express HLA-class II proteins. Many times, the same or similar peptide antigens are expressed through both the class I and class II systems. This allows both types of T cells (CD4 and CD8) to participate in the response.

Many immunogenic viral peptide antigens are expressed this way, which results in a coordinated immune response. Often, the virus-infected target cell is killed before it releases infectious viruses. This limits the ability of the virus to multiply. Another difference between the class I and class II systems is that the peptide seen by the CD4 T cell is processed through an exogenous (external) pathway. This pathway uses the ability of APCs to engulf (phagocytize) cellular debris into "cellular bags" (endosomes) and reduce the proteins inside into smaller subunits (peptides) through a series of steps that involve enzymes (proteolytic) that break apart the ingested protein material. These peptides are usually 8–12 amino acids long. The peptides are then associated with the HLA-DR molecules in another series of steps prior to their expression on the cell surface of the APC. This will be covered in much more detail as it relates to how changes in this process can profoundly change the outcome of the immune response. Immunologist Jonathan Sprent (1999) discusses the role of MHC class II proteins in the activation of naïve T cells.

The T cell Response to Presented Antigens

The previously mentioned combined CD4/CD8 response is based on the initial response of the CD4 T cell to the HLA-DR-non-self peptide antigen combination. These initial stages of the immune response are critical to the outcome of this battle between our immune system and tumors, viruses, bacteria, and other pathogens. A successful immune response is one that eliminates the non-self threat and leaves behind a memory response to ensure that future outbreaks do not occur. This response relies heavily on the first few encounters between the immune system and the threat. The T cell immune system can either work for the host or against it. It could mean the difference between life and death of the host (you and me). Many of the initial decisions of the early immune response occur in the class II or HLA-D antigen presentation system. This response is dependent on which of the 8–12 amino acid peptides are selected to be associated with the HLA-D protein on the surface of the APC. Apparently, the wrong or inappropriate peptide can be expressed, resulting in the generation of an immune-suppressive or anti-self immune response (Edwards 2010).

Autoimmunity and Immunosuppression

As mentioned above, the thymus has the ability to fine-tune the T cell response to non-self antigens by eliminating self-reactive T cells. However, some self-reactive cells are not eliminated. The reason is that some of these T cells are needed to control or eliminate T cells during an immune response that has become harmful to the host. For instance, during an immune response, T cells proliferate at a tremendous rate. Sometimes, somatic mutations (errors) appear in the genes of the TCR. These errors can change the cellular target under T cell attack. For instance, the flu pandemic in the 1900s resulted in the deaths of millions of people. Australian scientists Peter Doherty and Rolf Zingernagel discovered the nature of the HLA antigens and described how cytotoxic T cells attacked and killed flu virus–infected cells. They also suggested that the millions of deaths during the pandemic were not actually caused by the virus itself, but were the result of T cells attacking normal cells. The virus induced such a strong immune response that the T cells possibly lost the ability to distinguish between self and non-self. Therefore, influenza-activated T cells attacked normal tissues, which resulted in many deaths. This type of autoimmune reaction is very serious. It is feared that one day such an epidemic will happen again. Scientists are carefully watching the "bird flu". They are hoping it will not recombine with human DNA and cause another pandemic. Doherty and Kelso (2008) discuss the flu virus and the development of broad-based vaccines that can protect people from new versions of the flu.

Suppose the target of a T cell is a tumor cell. A genetic TCR error might change the target to a normal cell from which the tumor had developed. This results in an autoimmune response that can be deadly. Other T cells (T regulator cells or Tregs) have the ability to stop these "renegade" T cells. They begin a series of actions that directly affects the anti-self T cells, but also may slow down and sometimes retard the ongoing specific (tumor) immune response. Regulatory T cells can induce death of the renegade

T cells by activating "suicide genes" in the cells but also can kill them directly. In addition, they can initiate other regulatory measures, such as the elaboration of immunosuppressive lymphokines, to stop the renegade T cells.

This cascade of events can be triggered not only by the presence of renegade T cells, but also, as mentioned above, by the presentation of inappropriate or degenerate peptides. In this case, the CD4 TCR can induce a negative signal to the cell if it sees either a partial (not complete sequence) peptide or a series of sequence substitutions in the presented peptide. This immune-suppressive response is an active one where T cells that are causing it are continually dividing and reproducing. This event shuts down all productive responses and leaves the immune system unable to respond to not only simple infections, but also complex viruses and malignancies. If this cascade of events is triggered during that initial immune response, every response after that works against the host and for the invader. Viral infections have been shown to immediately induce these responses and they do it through a combination of specific actions on the antigen presentation pathway (Norris et al. 2006). This will be examined in the section that describes antigen presentation of viral antigens.

How Antigen Presentation Works at a Protein Level

The CD4 TCR is designed to have an affinity for a peptide, which is sitting inside what is called the HLA-DR "binding groove". The peptide remains in place through a series of covalently (charged) linked contact points between it and the amino acids that make up the peptide binding region of the HLA-DR alpha and beta proteins. The amino acid arrangements of the binding region vary quite a bit. This means that there are many different peptides that can fit into these grooves and subsequently be expressed on the cell surface for multiple TCRs to recognize. In addition, this cell surface combination is in flux. These cell surface complexes are continually being internalized and the peptides removed in acidified endosomes (Levine and Chain 1991).

As important as the surface expression of these complexes is, what happens inside the cell during their formation can affect the very nature of the immune response to them. Many molecular decisions are made. There are multiple steps involved after the synthesis of the HLA-DR proteins. There is the association and disassociation with chaperone and peptide editing proteins, peptide binding, and endosomal fusion with the outside cell membrane. This leads to their eventual arrival on the APC surface loaded with immunogenic peptide. In addition, there are many more complex molecular interactions that are based more on genetic changes than on biochemical reactions. As complicated as this process seems to be, it is a routine occurrence and is the basis of the ability of our immune systems to function normally. Again, without this process, we would be defenseless against infections and malignancies.

How Antigen Presentation Proteins are Assembled in the Cell

HLA-DR alpha and beta proteins are constructed in the cytoplasm (interior of the cell). They exit into the endosomal compartments. After reaching these compartments, a protein (chaperone protein) is bound to the peptide-binding region. This protein is called invariant chain (Ii). It aligns itself with the binding groove, leaving significant amounts of protein trailing each end of the HLA-DR molecule (Neumann et al. 2005). These trailing elements guide the DR molecule to its next destination. Once this protein complex reaches the next compartment, enzymes (proteolytic) degrade the trailing edges of the Ii protein. What's left in the peptide-binding groove is called the invariant chain class II associated peptide or CLIP. CLIP is one-third the size of Ii. It is much like the peptides destined to occupy the binding groove. However, it is a very special peptide. Its sequence always includes amino acids that have the ability to occupy the binding grooves of any HLA-DR protein regardless of HLA-DR type genetic changes. This is quite an achievement since there are hundreds of HLA-DR types and another hundred genetic changes. Despite this, the complex moves to the next stage.

That next stage of antigen processing is the peptide-editing phase. It involves another member of the HLA-D family called HLA-DM. It's structurally similar to the HLA-D proteins in that there is an alpha and a beta chain. However, that's where the similarity ends. This protein (HLA-DM) is actually viewed as an enzyme. How can that be? Well, an enzyme is a protein that controls a biochemical reaction. HLA-DM does just that. This molecule attaches itself to the HLA-DR-CLIP complex. Once attached, it has the ability to release the CLIP and allow an immunogenic peptide to occupy the peptide-binding groove. The strange part of this process is that the HLA-DM molecule has the ability to choose which peptide is inserted into the groove. It has to do with the affinity or binding strength of the peptide to the binding groove (Busch 1998).

As a part of this dynamic process, HLA-DM molecules are present during the re-entry of HLA-DR-peptide complexes into the cytoplasm (inside the cell) from the cell surface. Once the DR-peptide complexes are back inside the cell, the HLA-DM molecule controls the removal of the recycled peptide, then adds a new peptide from the milieu of peptides available to be presented by the APC. It's a continuum. If there is an invading virus or tumor present, its peptides get shuttled into the system. Which peptides get presented and what factor the invader (virus or tumor) presents to subvert that choice are the variables associated with the immune response against them. Rocha and Neefjes (2008) review the various issues involved with MHC class II proteins and peptide antigen presentation.

Genetic Changes that Affect Antigen Presentation

There is evidence that some genetic inherited changes can predispose some of us to get a disease (viral or cancer). As far as antigen presentation is concerned, there is a plethora of information that says inherited genetic changes affect the way our immune systems fight infections or malignancies. Minor changes that occur in our DNA can potentially make us unable to respond to infections or malignancies. An example is viral and bacterial meningitis. Usually, young people get it and they are often in high school or college. Local health officials immunize people who have come in contact with affected persons. Almost never is there any transmission of the bacteria or virus. Why is that? Well, because the bacteria or virus is always present in our environment. The problem is that there are some individuals who have a "hole" in their immune system (see Chapter 15). In other words, the immune system is unable to see viral or bacterial peptides and respond to them. People without this hole in their immune system immediately see these peptides and eliminate the cause of the meningitis easily. The source of this difference is a genetic mutation in the immune system. It probably resides in the genes that make the T cell receptor or possibly in APCs. However, this mutation helps us survive as a population, since not all of us are infected by the same pathogen. Ideally, the survivors are supposed to come up with a treatment.

Why Antigen Presentation of Viral Antigens Is Important

Viruses seem inanimate compared to cells, which are living, breathing entities. Cells are miniature factories that produce protein products and transfer oxygen and electrical signals that allow us (the host) to benefit from their efforts. Viruses may seem lifeless, but they are very clever at what they do. They enter a cell and take over the factory. The cell, instead of making useful proteins, is now commandeered to assemble copies of the invading virus. This process of infection, takeover, and replication is complex. Each virus has its own variation of these events. There are thousands of viruses. Many viruses attack specific cellular targets. However, their objective is the same. Once inside a cell, a virus makes copies of itself, ruptures the cell, and releases these copies to re-infect any remaining cell not already infected. It can do this until the host dies or continues to infect others. Some viruses are very quick at this and some take very long. Once the viral program begins, the host must fight back or succumb to the viral infection. Over millennia, viruses have learned how to subvert the host immune system to the point where it is very difficult for the host to completely eradicate the infection. A key example is the common cold virus, which has completely eluded attempts by scientists to cure or stop the ability of this simple virus to cause the problems associated with the infection.

Hepatitis-C virus

Evidence shows that the outcome of infection with the hepatitis-C virus is related to the genetic make-up of the individual who gets infected. There are usually two outcomes. In the first, a person is infected (liver cells) and suffers from a chronic form of the disease. This often leads to liver cirrhosis, then liver cancer. At this point a transplant is warranted. In the second outcome, a person gets infected and removes the virus from their system. The reasons for this disparity are probably genetic and focused on the genes of the immune system. Of particular interest are the two genes that control immunogenic peptide selection during the antigen-processing phase of class II antigen presentation. They are the genes that code for the HLA-DR peptide binding groove proteins and HLA-DM protein.

If we examine the immunological characteristics of these two patient groups (chronic and recovered), it appears that the chronically infected patients have suffered from virus-induced immune suppression. On the other hand, the recovery group seem to have had a productive immune response to the virus and subsequently removed it from their systems (Kaplan et al. 2008). As previously mentioned, immune suppression can be induced by the presentation of antagonistic (inappropriate) peptides to CD-4 T cells during the initial phases of the response. This may be happening to people who get a chronic form of the infection. Peptide selection is mainly controlled by the protein configuration of the HLA-DR binding groove. The HLA-DM protein plays a part by being able to "edit" or choose which immune peptide gets placed in the binding groove. DNA sequencing of the genes responsible for the proteins that control peptide selection within these patient groups may reveal genetic changes, mutations (or lack thereof) that are associated with the clinical outcome of the hepatitis-C infection (Koike et al. 2004).

Herpes simplex virus

Herpes simplex virus (HSV) is indeed one of the most intelligent viruses when it comes to the immune system. Like many viruses, it attacks the very cells that would be responsible for its removal. The virus attaches itself to the cells using a viral protein that links to the cell surface. Virus-cell fusion occurs and the virus enters the cell. Unfortunately, this cell is a macrophage or other APC that is basically defenseless against such an attack. The viral protein now goes to work. It has been shown to bind to the HLA-DM protein. In addition, it can insert itself into the peptide binding groove. It inhibits immunogenic peptides from binding. It also slows down the process. HLA-DR-peptide molecules cannot exit the endosomal compartments with the same speed. Actually, this is a good thing for the virus, because by limiting the ability of the APC to express HLA-DR, it limits its ability to present viral antigens. This virus interferes with the pathway in such a way as to paralyze the immune system. The only problem is the virus cannot actually propagate inside the APC because of the interference caused by the insertion of the viral protein. However, there are many more cells that can be infected.

Leishmania bacteria

Some bacteria are as efficient as viruses in subverting the immune system (Neumann et al. 2003). Leishmania bacteria can cause serious problems. Leishmania is a water-borne infection common in Third World countries. The bacterium can produce a peptide that has the ability to suppress the immune system. It's an antagonistic peptide. Once presented by the APC, it induces an active immunosuppressive response. This is not good for the host. T cells begin to initiate the immune suppression cascade and that causes a meltdown of the immune system. This action allows the bacteria to flourish. This immune response to leishmania is discussed by Kamala and Nanda (2009) as it relates to antigen presentation/processing and HLA-DM.

Antigen Presentation and the T cell Immune Response to Cancer

Tumor cells have their own ability to fight the immune system so it would be pointless to leave them out of this discussion. Every cell in our body has the ability to become malignant. Once that occurs, the immune system has a problem: just before that cell became a cancer it was a normal cell. The immune system really has a hard time figuring this out. Fortunately, in some cases the tumor cell begins to express HLA-DR antigens. As pointed out before, HLA-DR expression may lead to tumor antigen presentation. This occurs mostly in early tumors (melanomas, prostate tumors, and others) that have the ability to present tumor-associated antigens. These antigens (peptides) are seen by the immune system as non-self. The immune system takes over, and the tumor cells are destroyed. This happens more often than you might think.

Tumors do have the ability to circumvent the immune system. What they do is present an altered self-protein. The immune system gets confused. Is it self or non-self? As mentioned above, this self/non-self confusion can take two forms. The first is the presentation of an altered peptide antigen, which can induce an immunosuppressive response. The second is an actual autoimmune response that leads to the destruction of normal cells. Often, these normal cells happen to be progenitors of the cancer itself. For instance, when some melanoma patients respond to their tumors, this confusion occurs. Normal pigment-producing cells (melanocytes) are destroyed, resulting in vitiligo, or loss of skin pigmentation (Gui et al. 2007). Other cells closely related to melanocytes, such as cells of the nervous system, are also attacked, causing a painful condition called peripheral neuropathy (Visovsky et al. 2005). During these episodes, the immune system's attention is focused not only on these two responses, but also on how to stop them through the recruitment of regulatory T cells that are designed to shut down such responses. This confusion allows the tumor cells to multiply. After some time, the cells begin to move (metastasize) inside the body. The tumor cells end up in the liver, lungs, skin, and finally the brain. Then the host dies.

How New Advances in Antigen Presentation Research Can Affect the Future of Cancer Research

Let's focus on some of the new antigen presentation research that may lead to the development of novel immunotherapeutic treatments. As previously mentioned, gene therapy or genetic transfer of information can be a powerful tool. Turning cells such as tumor cells, which are usually unable to present tumor antigens, into effective APCs can give the immune system a chance to respond. As an example, anti-tumor cell immune response can be initiated by the injection of genetically altered non-dividing tumor cells from the patient. It can target existing non-altered tumor cells throughout the body. Another variation of gene transfer is the fusion of an APC with a tumor cell. This allows the APC's processing and presenting machinery to present the tumor antigens. These fusion hybrids have been shown to induce immunity to previously non-immunogenic tumor cells.

Another innovation is DNA vaccination, which is really not new. Pioneered in the early 1990s, DNA vaccination is the process by which DNA segments of vaccine proteins are inserted in the skin and muscle. The DNA immediately begins to make proteins that the immune system can respond to. What's new about this process is the protein made by the DNA. It is a protein that increases the ability of APCs to engulf immunogenic proteins and expose them to the endocytic HLA-DR-peptide pathway.

Acceleration of this process has been shown to magnify the T cell response (Nchinda et al. 2008).

One interesting development is the discovery of synthetic small molecules that can allow the HLA-DM enzyme to more rapidly exchange CLIP for peptides and new peptides for old ones. The end result is increased appearance on the cell surface of immunogenic HLA-DR-peptide complexes. This increases the likelihood of a T cell response because the APC is sampling many more T cell receptors. When the proper T cell receptor is found, the CD-4 T cell begins the immune cascade (Marin-Esteban et al. 2004).

However, with all of these new developments, experiments must include controls that look at immune suppression and autoimmune reactions. As mentioned earlier, these two conditions could lead to diseases just as bad as what the original immune response was fighting, or worse. The use of these new methods may be able to lessen the strength and duration of chemotherapy. This most certainly will reduce the harmful side effects of these agents. These advances have the ability to modify the T cell response to cancer in ways that may make it easier to generate additional cancer antigens (by antigen processing) that could be targets for other T cell effector cells not usually included in the anti-cancer T cell response. This occurrence would have a profound effect on the nature and potency of the T cell attack.

The Future

On the basis of years of research and technological advances, the trend now is to develop therapies that do not rely on toxic drugs, but involve the manipulation of our immune systems. Treatments like gene therapy and transfer of pre-conditioned immune T cells are just the beginning of a less toxic treatment for malignancies and viruses. Scientists may use the research of others to fuel their imagination. Ideas spring from this knowledge, a powerful tool. This tool has brought about the advancement of mankind for thousands of years. However, there is the problem of connections between science, politics and money. The pharmaceutical industry is an example of for-profit science. The industry makes drugs. Some of these drugs have extended the lives of people around the world. The companies do a good job and the cost of developing new drugs is very high. Sometimes our government wants such large corporations to flourish. This may make the Food and Drug Administration lax in allowing drugs that have not been examined thoroughly to come into the market. Occasionally, the side effects of these drugs have resulted in the deaths of many in this country. Non-profit science has been traditionally performed at universities where scientists pursue knowledge for its own sake. This is very different from a company-scientist pursuing knowledge for the development of drug treatments and profit.

These thoughts, this knowledge of the immune system, are rudimentary at best. They describe a basic function that, though seemingly routine, is the basis of our existence as a species. There is much more to learn, especially from other species. Sharks, for instance, never get cancer. Their immune systems are thousands of years older than ours. They survive and live through the same chemical and biological hazards that we do, but they do not get sick. Our survival depends on the survival of species much less developed than we are. We can think, reason, and be aware of our surroundings. Does that make us superior? I don't think so. We have to protect all life even if it seems hostile to us. Already, thousands of species of plants and animals have disappeared, some due to evolution, but many because of our own growth as a species. This leaves us with less and less knowledge to draw from.

One exciting aspect of immunology is the development of stem cells. These primordial cells can be directed to become any kind of cell in the body. This can start a brand new frontier of research in immunology. Actually being able to create T cells from a patient's own tissue brings hope to the millions of people suffering from malignancies and viral diseases. In time, a type of immunotherapy that has no side effects will replace the use of chemotherapeutic agents. This could potentially leave patients with an immune system that can fight back against a disease that would have killed them. However, right now we must depend on the efforts of medical science to unlock the mysteries of antigen presentation. How we are able to manipulate this system and how we incorporate those manipulations into the treatment of disease will determine our fate as a species.

Chapter 3

T cells and Cancer Cells: Components of Immune-based Anti-cancer Therapy

Chapter 3 contains an overview of the conflict between the immune system's T cells and cancer cells that develop in our bodies. It is a critical battle that can often determine whether we can survive the diagnosis of cancer. Learning about what factors determine the outcome of the battle between these two cells will enhance our ability to understand how new immune-based anti-cancer therapies are developed. The therapies will be discussed in detail later, but they make use of our own immune systems. Therapies like this minimize or eliminate the use of chemotherapy, which often has unwanted side effects. In an effort to understand how immune-based therapy works, it is essential that we learn about the two cells involved in its development. After this, we describe how the immune system works together to help T cells attack the cancer. Next, we examine the ways in which cancer cells cause problems by circumventing the ability of T cells to eliminate them. We also examine the definition and history of immunotherapy as it relates to current immune-based therapies. Finally, we review new experimental approaches to T cell-based immunotherapy. This will give us an idea of what the future of immunotherapy will look like. The information developed today will determine whether the T cell wins the battle against the cancer cell.

The T cell

The T cell, as well as other white blood cells, is born in the bone marrow. The new T cell is actually quite small. It is slightly larger than a red blood cell. Ninety percent of the cell is nucleus and there is very little cellular material (cytoplasm). The newborn T cells go directly to the thymus, a small gland in the throat where the T cells are conditioned to enter the rest of the body. Several things happen to the T cells during this time. First, the T cells are separated into two or more groups. The two major groups are the immature helper T cell (CD4) and the cytotoxic T cell (CD8). Each type of cell possesses a T cell receptor (TCR). Once this segregation occurs, the cells are screened through the TCR for their ability to respond to self. If they do respond to self, they are eliminated. In the end, what remains are T cells that have the ability to respond to mostly, but not always, non-self. We will explain this later when we discuss autoimmunity.

The T cells now leave the thymus and enter the bloodstream. Once in the bloodstream, they can, if summoned, go into the tissues, initially traveling through single cell–sized capillaries until they encounter solid tissue. The solid tissue actually does not stop them. T cells are so small and amoeba-like that they can squeeze through cellular junctions to their destination. After being summoned to their destination by proteins secreted by other T cells at the site, they begin to search out and respond to non-self proteins. Non-self proteins include viral, bacterial, fungal, and tumor proteins. However, there are special circumstances that allow for this response to happen.

T cells can respond to non-self protein only if they are processed into peptides by an antigen-presenting cell (APC). These cells will be described later. Processed peptides get expressed on the cell surface of the APC as a complex that consists of the non-self peptide and a self-protein (HLA-DR). This complex is what is seen by the TCR. Once this receptor is engaged, a series of events occur inside the T cell allowing it to respond. This is called activation. As mentioned above, generally two types of T cells (T cell subsets CD4 and CD8) participate fully during an immune response. Each type responds differently and plays a separate role in the T cell immune response (Feldmann et al. 1989).

The T cell Subsets

The CD4 T cell is always the first responder. After it becomes activated, it secretes a host of small proteins called lymphokines. These proteins target other cells of the immune system to instruct them how to respond. One of the most important proteins is called Interleukine-2, or IL-2. When the CD8 T cell sees the non-self/self-protein (HLA), it needs to see the IL-2 molecule in order to be activated. CD8 activation and killing of its target will be discussed later. The role of the CD4 T cell in T cell immunity is pivotal and vital. The combination of lymphokines secreted by these cells can either stimulate the immune response or shut it down. All that depends on the type of signals received by the CD4 T cell during the activation phase. Once an immune response is started, it sometimes has to be shut off because of harmful side effects to the host, such as prolonged fever and destruction of normal cells. The destruction of normal cells is called autoimmunity.

The CD8 T cell response is different from that of the CD4 T cell. First, the CD8 T cell sees non-self associated with an HLA protein, whereas the CD4 T cell sees non-self associated with HLA-DR proteins. The major function of the CD8 T cell is to kill other cells that are either infected with viruses or are cancerous. It can do this in several ways. It can poke a hole into its target cell, which lets out the contents of the cell, and the cell then dies. Or it can activate a suicide pathway in its target, which also kills it. These types of cell death are important because they can induce different types of immunity. CD8 T cells are the enforcers of the immune system. However, they have another role and that is to control other T cells that can cause harm to the body. As mentioned above, the response to self is called autoimmunity and it happens despite the elimination of anti-self T cells in the thymus.

A classic instance of autoimmunity on a large scale occurred during the flu pandemic in 1918. It was initially thought that the virus caused the deaths of millions of people. But it was determined that it was not just the virus but a side effect of the infection that caused so many deaths. The immune response to the virus was so strong that the CD8 T cells lost the ability to distinguish between self and non-self and subsequently began to attack normal tissues. Perhaps today this type of autoimmunity could be addressed medically, but back then no one even knew it was happening. It took the Nobel Prize–winning work of two scientists from Australia (Peter Doherty and Roff M. Zinkernagel) to uncover this phenomenon.

The ability of T cells to control our immune systems is crucial. The T cells that control anti-self T cell responses are called T regulatory cells (Tregs). Tregs are derived from the CD4 subset. They have specialized genes activated to produce proteins that control the actions of other T cells. Without this control we would suffer from diseases such as lupus, arthritis, and several other autoimmune diseases at a much higher rate.

The Cancer

Carcinogens, viruses, and other environmental factors cause breaks in the DNA of normal cells. The DNA tries to repair itself, but mistakes are made. These mistakes are called mutations. When they occur in genes that control cell growth, the cell begins to divide and reproduce itself in an uncontrolled manner. This is the simple version of how cancer starts. The DNA in our cells is the key. Controlling the growth and spread of cancer cells, for the most part, is the responsibility of our immune systems, particularly the T cell. Almost each cell in our body has the ability to become cancerous. This causes multiple problems within the immune system. First, before a cell becomes cancerous it was a normal cell. This problem will be revisited during the discussion of how cancer can subvert the immune system.

Second, as the cancer cell continues to grow unabated, it begins to use up resources at a rapid rate. This is a threat to the host. If left untreated, the cancer will eventually kill the person who has it. However, our immune system has a failsafe mechanism and that operates more often than we think.

As cells in the body become cancerous they start to express the non-self/self-protein molecule complexes and present tumor antigens just as an APC does. CD4 T cells immediately begin to respond to the cancer. They also do something special; they secrete a lymphokine called gamma-interferon. Gamma-interferon enters the cancer cell and causes a ten-fold increase in expression of this complex. This induces even more CD4 participation. Ultimately, CD8 T cells are called in by IL-2 and begin to kill the cancer cells. Why cells do this before and during the process of becoming cancer cells is still not quite understood. It is a way for these cells to alert the immune system that a cancer is forming. It probably happens often, as mentioned above, but is hard to measure because so few cells are involved. It becomes a problem only when the transforming cell grows rapidly or does not express the non-self/self complex at all. Both scenarios have caused the death of millions of people (Yigit et al. 2010).

There are many different types of cancers, just as there are many different cells in our bodies. Generally, they are classified into two groups. The first are the solid cancers that originate in a specific location and begin to grow there, such as melanoma, brain tumors, and liver cancer. The second are the blood-borne cancers, which originate from cells in the bloodstream such as white blood cells. Examples of these cancers are leukemia, lymphoma, and myeloma. Each group is treated differently. For instance, both groups get treated with traditional chemotherapy, but the blood-borne cancers need more aggressive treatment because they, like the white blood cells they come from, originate in the bone marrow. Often, chemotherapy can remove them from the circulation, but the bone marrow just creates more cancer cells over time. The cancer can be eliminated by a bone marrow transplant from a sibling or tissue type–matched donor. Therapy of this type sometimes involves the manipulation of the immune system by anti-rejection drugs. Recent research has led to the transplantation of donor cells that have had the immune cells modified in order to eliminate a life-threatening condition associated with bone marrow transplant. We will discuss these modifications when we explain new immune-based anti-cancer therapies.

What You Need to Know about the Rest of the Immune System

The immune system is continually watching the body for the presence of a threat. This threat can come in many forms: bacteria, viruses, other pathogens, and, of course, cancer cells. This is called "immune surveillance", the continual search for non-self. Normal cells in the body are recognized as "self" during this process. However, when these cells become cancerous, they produce proteins that the immune system cannot recognize as self. The surveillance system now begins to take notice. Three types of immune cells are involved in this continual activity. They are the natural killer (NK) cell, T cells, and the APC. The APC is usually termed a macrophage, but there are many variations of this cell type. The NK cell has the ability to recognize an irregular cell surface associated with the tumor cell and then promptly bore a hole in the cell, killing it. The killing of tumor cells by NK cells is rather indiscriminate and short lived, but the initial death of some tumor cells is very important (Berzofsky and Terabe 2009). It is important because the APC will ingest the dead tumor cells, find the non-self proteins, process them into peptides, and present them to CD4 T cells (Stout et al. 2009). As mentioned before, a T cell cannot respond to non-self unless it sees non-self associated with self. The APC links non-self peptides to self-proteins and they get put on the APC surface where CD4 T cells orchestrate the T cell immune response.

Problems Associated with the T cell Immune Response to Cancer

For T cells to properly respond to cancer, they must overcome autoimmunity and T cell suppression of an immune response. In cases of autoimmunity (Juszczak 2009), the cancer cell can sometimes subvert the immune system by making it think it is responding to normal cells. For instance, when some melanoma patients respond to their cancers this dilemma occurs. Melanoma cancers originate from normal pigment-producing cells called melanocytes. During the T cell immune response to melanoma, some of the T cells attack normal melanocytes because of the similarity between them and the melanoma. When melanocytes are destroyed, there is loss of pigmentation of the skin. This condition is known as vitiligo (Fujimoto et al. 2006). Other cells closely related to melanocytes, such as cells of the nervous system, are also attacked, causing a painful condition called peripheral neuropathy. During these episodes, the immune system's attention is focused not only on these two responses, but also on how to stop them through the recruitment of regulatory T cells that are designed to shut down such responses. This confusion allows the cancer cells to multiply.

Induction of cancer-specific T cell immune suppression is the other major problem. During the process that turns a normal cell into a cancer, the proteins become mutated or simply cut in half. Unfortunately, when a T cell sees these proteins, its natural response is to start a shutdown of the T cell immune response associated with that cell expressing the target protein. In other words, any T cell, regardless of the type or stage of activation, is arrested. The T cells are prevented from responding. This failsafe mechanism programmed into T cells allows them to pause an immune response. Normally, this mechanism is used to prevent a T cell response to subtle changes in normal cells or the emergence of fetal antigens during pregnancy. This is another way a cancer cell can subvert the immune system and grow seemingly unnoticed while the immune system is paralyzed. Some of these mechanisms will be covered in detail later because these problems encountered by the T cell immune system during an anti-cancer T cell response are at the heart of its failure to eliminate the cancer. After these problems are highlighted in the chapters that follow, the hope is that a new appreciation for the complex nature of the anti-cancer T cell response will become evident.

Immune-based Therapy for Cancer Treatment, Its History and Use to Fight Leukemia

Immune-based therapy or immunotherapy uses the cells of the immune system instead of chemicals or radiology to fight cancer. It makes use of the extensive knowledge of the immune system accumulated over the last 50 years or so. With this knowledge, scientists can manipulate the components of the immune system to target the cancer more specifically. The components of the immune system range from those small proteins called lymphokines to genetically altered or re-programmed APCs. T cells can also be re-programmed to attack cancer cells expressing specific proteins. T cells are probably the most important component in immunotherapy because they ultimately kill the cancer. However, the use of other immunological factors is essential to the generation of a productive anti-tumor T cell immune response.

Now let's look at the history of immune-based therapy. This type of treatment began in the 1970s while scientists were studying a mouse tumor model. The tumor was blood-borne and usually lethal, but they found an active immune response to it. The tumor cells were grown in the lab. Once they were inactivated, they were cultured with normal mouse T cells over a series of days in a temperature- and gas-controlled incubator. After this culture period, it was found that the T cells had become immune to the tumor with the ability to kill them. Subsequently, massive amounts of these T cells were produced. Mice were given an inoculation of small numbers of live tumor cells followed by an injection of tumor-immune T cells. Under controlled conditions, it was discovered that the mice that received the immune T cells survived much longer than those that did not. This technique was called "adoptive transfer". It ushered in an entirely new approach to treating cancer (June 2007).

Leukemia, as we have said, originates from the bone marrow, which is the birthplace of both white and red blood cells. Unfortunately, the people who most often get this disease are young children and the elderly. Some forms of leukemia are curable, especially in young patients. The treatments can be painful and have many side effects. At first, chemotherapy is used. If that fails, a bone marrow transplant is the next step. Since the leukemia cells originate in the bone marrow, it is necessary to remove all the cells in it and replace the marrow itself. The donors of the bone marrow are often, but not always, matched to the genetic type of the recipient. The bone marrow of the patient must be completely destroyed before the donor marrow is transferred. This treatment involves a combination of chemotherapy and radiation. Shortly thereafter, the patient is devoid of white blood cells and thus susceptible to life-threatening infections. The donor marrow is infused into the patient and the cells begin to reside in the patient's marrow.

Even if the donor is an identical genetic match to the patient, there is a possibility that the donated marrow cells will begin to attack the patient's tissues.

This is called graft vs. host disease (GVHD) and can be deadly (Barton-Burke et al. 2008). At present, the patient must be given anti-rejection drugs until the grafted cells establish themselves. These drugs have harmful side effects, such as immunosuppression, which can again lead to deadly infections. However, research is being done to identify the T cells that mediate this process and remove them prior to the transplant. These T cells see very slight genetic differences between the donor and the recipient even if the two individuals are twins. The genetic changes are due to minor changes that occur in our DNA after we are born. The donor T cells see the protein products of this DNA as non-self. This triggers a response and tissues with these proteins are attacked and damaged. The tissue that gets damaged can be a vital organ, such as the intestine, liver, or skin. Once scientists identify the protein targets of the tissues under attack, it will be simple to remove those T cells from the donor's marrow. It will reduce or eliminate the life-threatening effects of GVHD. Refinements to techniques like this one will minimize or eliminate the use of anti-rejection drugs and the side effects associated with them.

More Recent Advances in Anti-cancer T cell Immunotherapy

Recent advances in genetic transfer have allowed scientists to genetically engineer T cells by inserting a manufactured T cell receptor that sees a non-self cancer peptide for that particular cancer. These genetically altered CD8 T cells go directly to the cancer site. At this time, the patient is administered the lymphokine IL-2 and the T cells mature into cytotoxic

T cells and kill the cancer cells. This technique has been shown to be effective in a significant number of patients who had advanced cancers. Indeed, some of them have been cancer-free for more than five years, which is the definition of cured (Zhao et al. 2005).

T regulatory cells, or Tregs, can play a major role in preventing GVHD because these specialized T cells can attack anti-self T cells that induce the disease. That is their natural function, as you will see in Chapter 11. In the past it has been difficult to isolate enough Tregs for this purpose. Recently, two drugs that add methyl groups to DNA have been shown to automatically convert non-Treg T cells into Tregs. Researchers have shown that these chemically induced Tregs inhibit the development of GVHD but allow the grafted T cells to attack the leukemia cells. This is a very recent advance and is still in the preliminary stages of development in mouse models of leukemia (Choi et al. 2010). The hope is that one day leukemia can be treated by bone marrow transplant without the deadly side effects of GVHD.

Within the last few years, anti-cancer vaccine approaches have been modified to specifically target cancer cells. For instance, we have said that APC could present cancer non-self/self complexes to T cells in order to induce immunity. Antigen-presenting cells from patients have been isolated, allowed to present specific cancer peptide complexes, and returned to the patient's circulation. The results have been very encouraging because there is a measurable immune response to the cancer and many of the cancer masses have been reduced in size.

Another variation of this concept involves the lymphokine gamma-interferon. As previously mentioned, this protein can increase the ability of tumor cells to display T cell–recognizable tumor proteins on the cell surface like an APC. The gene of this protein has been inserted into an APC and has been shown to accelerate the death of tumor cells by immune T cells. These studies are now being conducted on mice, but eventually they will become part of standard anti-cancer immunotherapy in humans. Antibodies have also been genetically engineered to attach to both the cancer and the APC. This forces APC processing and presentation of the cancer non-self/self complex to T cells (Valone et al. 1995). In addition, it has been shown that some of the new chemotherapeutic agents that kill tumor cells have the ability to render the dead cancer cells more easily processed by APC. That means that as the tumor dies the immune system is activated by what is now a vaccine. This leads to the destruction of even more cancer cells.

DNA vaccination, pioneered in the early 1990s, is the process by which DNA segments of vaccine proteins are inserted in skin and/or muscle. The DNA immediately begins to make proteins that the immune system can respond to. What's new about this process is the protein made by the DNA. It increases the ability of APCs to engulf immunogenic proteins and express them on the cell surface. Acceleration of this process has been shown to magnify the T cell response to cancer proteins originating from cancer cell death or an administration of cancer-specific vaccine (Wei et al. 2008).

Immunotherapy of blood-borne cancers has shown great promise. Recently, during bone marrow transplantation, it has been shown that if the donor of the marrow was not a perfect match for the patient and sophisticated anti-rejection drugs were used, the cancer could be cured. The reason for this is that the anti-rejection drugs prevent the donor's T cells from attacking the patient's normal tissue. Given enough time, these T cells disappear from the population, leaving only the anti-cancer T cells. Eventually the drugs can be withdrawn. As a result, these patients are cancer-free and have something most of us do not: two different populations of T cells. This is called chimerism (Bader et al. 2008).

In the next few years, cancer treatment should be painless and without side effects. New technologies are continually being developed to deal with this disease. Complete eradication of some types of cancer is a reality today, but just a few years ago it was a death sentence. Much more research must be done to eradicate cancer. The questions are how much research and how long will it take? Scientists seem to be very optimistic about the future of cancer treatment, but they also realize that it will take a major commitment by government and the private sector to solve the cancer problem.

Chapter 4

Anti-cancer Vaccines: New Approaches

Vaccines came about in the late 1940s and early 1950s. They were initially designed to combat bacterial infection. It was found that proteins produced by bacteria induced an immune reaction that was directed at the live bacteria in the event of infection. In the early days it was not known how this occurred. Vaccines were subsequently developed for polio, measles, and chicken pox. These vaccines were sterilized proteins injected usually in muscle tissue. Early evidence suggested that the cells in the blood somehow responded to the vaccine and developed the immunity to the pathogens. This immunity or "memory" of these pathogens was found to last some time. As time went on, it was found that white blood cells (T cells) were responsible for such immunity.

With this discovery came the realization that there could be a vaccine for cancer. However, cancer was different from pathogens because it originated from the body's own cells. It turns out that the initial vaccine experiments with cancer used the same mechanisms the immune system used to respond to pathogens. What was unknown was why the anti-cancer vaccines did not work the way the anti-pathogen vaccines worked. The anti-cancer vaccines induced a limited T cell response, which was unlike the anti-pathogen vaccine preparations.

It is this unknown that will be the subject of our discussion of how the old concept of vaccine use has given rise to new approaches to the development of anti-cancer vaccines. The new approaches to the development of anti-cancer vaccines have initiated major advances, which will revolutionize the effectiveness of vaccine therapy for cancer. In this chapter, we review how vaccines work and the evolution of their use so that we can understand how immunobiologists began to unlock the secrets of the immune system. Next, we explain how the T cell immune system responds to vaccine proteins while generating a protective immune state. We talk about why anti-cancer vaccines have to break through anti-autoimmune controls and how breaking this control could be harmful. We describe various types of new anti-cancer vaccine treatments in order to understand the future direction of vaccine therapy. Vaccine therapy, which uses the T cell immune system to protect us from bacteria, viruses and especially cancer, has the advantage of having fewer unwanted side effects.

How Vaccines Work and How They Have Changed

In a simple and early version of a vaccine, an attenuated (dead) organism, inactivated live organism, or protein is injected into the body to induce immunity against that organism. That organism may be either a bacterium or a virus that potentially could infect the individual. Using the organism example to explain how vaccines work will simplify the process of understanding the factors involved in the biological response caused by the vaccine. Anti-cancer vaccines, although based on these concepts, are more complex and will be discussed later.

The function of a vaccine is to induce immunity. However, it can be used as an active treatment to increase the immune response. The initial activity of the immune system on a cellular level is to process the proteins of the organism into recognizable subunits called peptides. In the first step, an antigen-presenting cell (APC) or macrophage engulfs the foreign proteins and then reduces them to small peptides. For the next cell to recognize the peptide, it must be linked to another protein that signifies self (peptide-self complex). This cell is called the T cell. At this point, two parts of the cellular immune system are generated to respond to the organism from which the protein came. The first, and the focus of further discussions concerning vaccines, is the T cell. This cell has the ability to kill any cell expressing the peptide-self complex on its surface. The second is the antibody response, which is the result of T cells secreting proteins that cause B cells to make and secrete cancer-specific antibodies. Some of these antibodies can directly kill the cancer, but others attach themselves to the cancer and cause the APC to kill the cell. In fact, scientists first used antibodies to demonstrate immunity induced by vaccine treatment.

Vaccine preparations in the early stages of vaccine development were crude protein preparations. Sometimes, dead bacteria (usually killed by heat) were administered. Inactivation of a bacterium or virus was preferred because these preparations elicited a strong immune response. To test this response, scientists would take a small amount of the preparation and place it under the skin, usually 2 to 14 days after immunization. If the immunization was successful, a red spot would appear at the site, surrounded by a white circle with lines radiating from the center to the edge of the circle. This was called the "wheel and flare" response and was an easy way to detect immunity to the vaccine. However, it was noticed that some protein preparations were weak in inducing a response, so additional methods were used to increase their strength. For instance, when the smallpox vaccine was administered, scarring of the site was necessary to ensure an inflammatory response, which also included T cells drawn to the vaccine site. These scars also indicated which individuals had been immunized (Clark and Adinoff 1989).

As technology advanced, so did the nature of the vaccine preparations. Immunologists determined which amino acid sequences of immune-stimulating peptides were seen by T cells. This led to the development of a new generation of vaccine preparations. Coupled with the advent of recombinant DNA technology, it made it possible for scientists to manufacture synthetic vaccines in the lab. The development was important because it eliminated contaminants from the preparations, leading to safer inoculations.

New methods of vaccine delivery, such as the vaccine gun, were also developed. The vaccine gun delivers the vaccine under the skin through a high-pressure fluid pump. Another development is DNA vaccination, which can also be administered with the vaccine gun, injecting DNA programmed to produce vaccine proteins that enter the bloodstream. This topic is discussed in detail in the section that deals with new vaccine treatments.

Our understanding of how vaccines work and how they have developed over the years has gone through a fundamental change. This change is evident in the prevalence of this biological process in our everyday lives. Vaccinations are mandatory for a number of infectious agents and can be administered to children as well as adults. An explanation of how the T cell immune system responds to vaccines and what problems are encountered can help us understand how important vaccines are in our everyday lives.

How the T cell Immune System Responds to Vaccines and Creates Immunological Memory

T cells initially respond to protein subunits called peptides. Because there are two types of T cells involved in this process, it is necessary to describe their activities separately. The same two types of responding T cells work together to eliminate the target against which the vaccine was directed. This activity also involves the secretion of bioactive proteins, not only by these two types of T cells, but also by the APC. The development of T cell memory is a critical aspect of vaccine therapy because in most cases, with the exception of cancer and some viral infections, it is the sole reason the vaccine is used in the first place. All these elements of vaccine therapy must function in concert for the T cell immune system to either attack an existing cancer or infection or remember it, sometimes for the individual's lifetime.

The T cells involved

Two types of T cells (T cell subsets) participate in the T cell immune response to vaccines: CD4 and CD8 T cells. They are immature prior to their contact with the vaccine peptides, which are expressed on the surface of the APC or target cell. The peptides are associated with self-proteins called HLA-DR and HLA-ABC respectively (peptide-self complex). Once contact has been made, the T cells become activated through biochemical changes within the cell (see Chapter 9 on translational immunology). This activation allows the T cells to mature into responder (CD4) and killer (CD8) T cells. These two mature cells now control every aspect of the T cell response to vaccine proteins and the subsequent development of an immunological memory in the form of memory T cells. The memory T cells are derived from both the CD4 and CD8 T cell subsets. The response starts on a small scale, involving a few CD4 T cells and APC, but eventually can involve up to one-third of the total T cell population. The magnitude of the response depends on the strength or immunogenicity of the vaccine protein.

The CD4 T cell is always the first responder. After it becomes activated it begins to secrete small proteins called lymphokines. The proteins target other cells of the immune system to instruct them how to respond. The most important of these proteins is called IL-2. When the CD8 T cell sees the non-self/self-protein (HLA-ABC), it needs to see the IL-2 molecule in order to be activated to kill its target. The role of the CD4 T cell in T cell immunity is pivotal and vital. The combination of lymphokines secreted by these cells can either stimulate the immune response or shut it down, depending on the signals received by the CD4 T cell during the activation phase. Once an immune response is started it sometimes has to be shut off because of harmful side effects to the host, such as prolonged fever and destruction of normal cells. The destruction of normal cells is called autoimmunity (Juszczak 2009).

The CD8 T cell response is different from that of the CD4 T cell. First, the CD8 T cell sees non-self associated with an HLA-ABC protein, whereas CD4 T cells see non-self associated with HLA-DR proteins. The major function of the CD8 T cell is to kill other cells that are either infected with viruses or are cancerous. It can do this in several different ways. It can poke a hole into its target, which lets out the contents of the cell and the cell dies. Or it can activate a suicide pathway (Reich et al. 2008) in its target, thus killing it. These types of cell deaths are important because they can induce different types of immunity. CD8 T cells are the enforcers of the immune system. However, they have another role and that is to control other T cells that can cause harm to the body. This role is later discussed in detail.

The memory

Once the T cells have mounted a successful response, what has to be left behind is the immunological memory. The nature of this memory is embedded in the DNA of a small subset of both CD4 and CD8 T cells. The memory T cells actually de-differentiate: the cells revert to an immature state so that they can exist in the body for years. The signals that allow them to remain immature are not understood, but the development of these memory T cells is what helps us survive. They remain in our bodies for years or even our entire lifetime, untouched by aging or illnesses.

There are two theories about the development of memory T cells. The first is the "linear" theory. As naïve (new) T cells become activated by an antigen, part of this process includes the immediate creation of memory T cells after the generation of effector T cells and effector memory T cells. The second is the "parallel" theory, in which the result of naïve T cell activation is that memory T cells are created simultaneously with the two other T cell types mentioned above (Holmes et al. 2005). Either way, this change in T cell maturation allows us to respond to recurring cancers as well as to a plethora of bacteria and viruses that we have a natural immunity to, or have been immunized against using vaccine treatment, or have had prior exposure to.

One of the most exciting aspects of memory T cells is that they have very long telomeres. A telomere is a region of repetitive DNA at the end of a chromosome that protects the end of the chromosome from destruction. Telomere length directly indicates how long the cell will survive. The DNA elements are very long in memory T cells, allowing them to age much more slowly than most cells in the body. How this happens during the development of memory T cells is still under scientific debate, but the fact has been proven (Rosenberg et al. 2009).

One of the most important aspects of anti-cancer vaccine experiments is the use of anti-cancer T cells to treat cancers in the body. This is called adoptive therapy and entails infusing patients with their own previously immunized anti-cancer T cells. However, before this can be done, the cancer must be isolated and cultured so that the immunization can be done outside the body. Alternatively, immune T cells can be isolated from the cancer mass itself. The T cells isolated from the growing cancer are expanded and then used for the infusion. This infusion can be accompanied by the administration of an anti-cancer vaccine to bolster the effectiveness of the infused T cells. The adoptive transfer method has been viewed as laborious and possibly mistake-prone. Scientists are now beginning to understand that cancer patients already have anti-cancer T cells within their bodies, but these cells are unable to respond because an anti-autoimmunity mechanism prevents the response.

Why Anti-cancer Vaccines Have to Break through Anti-autoimmune T cell Controls

Most anti-cancer vaccine therapy is not preventative but is meant to stimulate and direct the T cell immune system to fight the cancer, a complex problem. The anti-cancer vaccine must be formulated to induce T cells to attack a specific cancer peptide target(s). However, some of these anti-cancer targets are similar to self. This is related to the very nature of how a cancer hides from the T cell immune system. It is sometimes related to the location of the cancer, such as the brain. The brain has a barrier that can present problems (Anirban et al. 2007). For the most part, cancer cells hide by pretending to be normal cells.

This comes into play when some melanoma patients respond to their cancer. Melanoma cancers originate from normal pigment-producing cells called melanocytes. During the T cell immune response to melanoma, some of the T cells attack normal melanocytes because of their similarity to the melanoma cancer. The normal melanocytes are destroyed, resulting in vitiligo, or loss of skin pigmentation (Gui et al. 2007). Other cells closely related to melanocytes, such as cells of the nervous system, are also attacked, causing a painful condition called peripheral neuropathy (Visovsky et al. 2005). This phenomenon is called cancer-induced autoimmunity (Galgani et al. 2009). During these episodes, the immune system's attention is focused not only on these two responses, but also on how to stop them.

Once anti-cancer T cells exhibit this type of autoimmune response, the immune system reacts by calling on a special type of T cell. These are regulatory T cells (Tregs), a subset of both CD4 and CD8 T cells. Tregs have certain genes activated that allow them to stop the activity of any T cell responding in an autoimmune fashion. Scientists have known about these cells for some time now (Chattopadhyay et al. 2005). But they cannot yet differentiate a Treg that controls an anti-pancreatic (insulin-producing cell) T cell response from a Treg that controls an anti-cancer T cell response that has autoimmune properties. At this point, stopping all Treg activity is unrealistic, because Tregs control a host of autoimmune responses that, if allowed to proceed, would certainly lead to the death of the individual. Here lies the problem and a solution is dependent on this identification. If Tregs that just inhibit anti-cancer T cells could be identified, a temporary removal of those Tregs would allow many more of the cancer cells to be destroyed. A vaccine against Tregs that inhibit anti-cancer T cells could induce a temporary removal of these cells if the vaccine itself were used in a limited fashion. The presence of Tregs that target anti-cancer T cells is a problem not only for the development of anti-cancer vaccines, but also for the use of adoptive immunotherapy techniques.

New Anti-cancer Vaccine Treatments and Why They Are Important

The development of vaccines to fight cancer entailed isolating cancer cells, killing them with immersion in water or detergents, and using these membrane preparations as inocula. Many vaccines did not work because they were too dilute in the blood and could not be used before they were removed. Vaccines were then placed under the skin. In these cases, many times the immune system was not alerted because of the similarity of cancer proteins to self-proteins. An adjuvant was tried next. An adjuvant is an agent that may stimulate the immune system and increase the response to a vaccine, without having any specific antigenic effect. Adjuvants accomplish the task by mimicking specific sets of evolutionarily conserved molecules that include liposomes, lipopolysaccharides, and components of bacterial cell walls. They are still used, but not as the sole means of delivering a vaccine (O'Hagan 2000).

As the knowledge of how the T cell immune system functions increased, so did the sophistication of anti-cancer vaccines. It was discovered that macrophages (an APC) played an important role in anti-cancer T cell immunity by presenting important cancer peptides to T cells, which would in turn initiate the anti-cancer T cell response. In developing vaccines for cancers in the brain, it was found that if macrophages were fed brain cancer proteins outside the body and then returned, an increased T cell response to the cancer could be measured. These are called cellular vaccines and have become even more effective. The technique makes use of knowledge gained about which immune proteins accelerate T cell activity during the response to a cancer. Two important immune proteins are secreted by CD4 T cells that increase anti-cancer T cell activity: IL-2, which matures CD8 T cells to become killer cells, and gamma-interferon, which can induce cancer cells to increase expression of self proteins recognized by CD4 cells. Both these genes are placed inside macrophages, and once the proteins are expressed, a measurable increase in anti-cancer T cell activity can be detected. Vaccines based on the immune cell are still in development but may someday become part of an integrated immunotherapeutic treatment for cancer.

In DNA vaccines, DNA segments of vaccine proteins are inserted in skin and/or muscle. The DNA immediately begins to make proteins that the immune system can respond to. What's new about this process is the protein made by the DNA. It increases the ability of APCs to engulf immunogenic proteins and express them on the cell surface. Acceleration of this process has been show to magnify the T cell response to cancer proteins originating from cancer cell death or an administration of a cancer-specific vaccine (Wei et al. 2008).

Another recent development is the use of hybrid proteins to facilitate a cancer-specific T cell response. The use of hybrid proteins may not be viewed as using a vaccine *per se*, but they are designed to have the same effect. That effect is to direct the T cell immune system to respond to cells expressing specific cancer-associated proteins. For example, hybrid proteins have been developed through the use of DNA recombinant technology that are bi-specific. One half of the protein is a portion of an antibody that is directed against a protein on breast cancer cells. The other half is a portion of an antibody directed to a macrophage (APC) protein (Fc receptor) that, when activated, will eat the cancer cell once contact is made (Valone et al. 1995). Since the function of the APC is to process and present cancer peptides to T cell immune system, bringing the cancer and the APC together makes sense. In fact, this technique shows great promise as an immunotherapeutic tool. Positive results have been reported in late-phase clinical trials.

Where Do We Go from Here?

In conclusion, it is quite clear that the development of anti-cancer vaccines has progressed in a steady fashion. It appears that they will shortly be part of an integrated T cell-based immunotherapeutic treatment for cancer. The knowledge gained over the last 50 years concerning the function of the T cell immune system has fueled this progress. There are still many questions regarding the efficient use of these vaccines. However, at the current pace of acquisition of immunological knowledge, it will not be long before anti-cancer vaccines are a standard approach to cancer treatment. Technological advances in other fields are bound to have an impact on the development of these vaccines.

These new approaches to the development of anti-cancer vaccines are derived from the basic concepts put forth in the early years when vaccines were used to fight bacterial and viral infections. We may soon be able to answer the fundamental question "Should anti-cancer vaccines be directed at the immune system instead of at cancer cells?" The nature of the answer may revolutionize the use of anti-cancer vaccines because it will involve the immune system controlling itself. It would be ideal if this control could be manipulated to boost the T cell immune response to a cancer. The cancer would be eliminated and memory T cells would remain to prevent recurrence.

Vaccines, as we have said, are an integral part of our lives. They protect us (young and old) from diseases that would devastate our population. In the near future, anti-cancer vaccines and integrated immunotherapeutic treatments will cure many forms of cancer. Above all, these treatments could be painless and free of the unwanted side effects associated with current cancer treatments. The future of cancer treatment will enter a new era. Cancer may no longer be a painful and deadly disease but just one of many diseases that are entirely cured or that allow patients to live a normal life.

Chapter 5

How and Why Do We Get Cancer and Can Immune-based Anti-cancer Therapy Save Us?

Almost all of us have known someone who has been diagnosed with cancer. This disease originates from the cells in our bodies. It is defined by the unregulated growth of previously normal cells. Just about every cell in our body can become cancerous. Rapid cell division, which is typical of cancer cells, is not in itself a particular problem. Liver cells have the ability to rapidly duplicate (regenerate) themselves under certain conditions. Cells of the immune system (T and B cells) can multiply quickly in response to infections (bacteria and viruses) that alert the body's defenses. But these two examples of rapid cell division are regulated by the body and occur only under special conditions. Cancer causes not only rapid cell division, but also unregulated cell growth. Cell growth without regulation is harmful to the body because it diverts resources normally needed for the body to survive.

In this chapter we look at some of the history related to cancer research. We also clarify some of the underlying reasons why we get cancer by looking at the agents or conditions that lead to its development. In addition, we examine the relationship our immune system has with cancer and how it goes about the job of removing it. We look at how the cancer can undermine the immune system's effort to eliminate it. Because this discussion is about cancer treatment, we introduce the development of chemotherapy. We describe the advent of immunotherapy and discuss how this concept of immune-based cancer therapy is applied to human cancer. Also, we examine the combination of chemotherapy and immunotherapy. Finally, we discuss the future of immunotherapy as it relates to cancer treatment and its impact on our society.

History of Cancer Research

In 1970, cancer rates and mortality (deaths) were at an all-time high. Health care professionals and the public pressured the government to do something. President Richard M. Nixon declared a "War on Cancer" (Davis 2007) and signed the National Cancer Act. As a result, Congress allocated $100 million to the National Cancer Institute. In addition, the Frederick Cancer Research and Development Institute was created in Frederick, Maryland. It housed laboratory facilities for cancer research. Labs were also added to the National Institutes of Health (NIH) in Bethesda, Maryland. The NIH also became a grant funding institution, providing university scientists funds to research this disease. The funds were available not only to experienced scientists, but also to students and new Ph.D.s in the form of training grants. The training grants provide new and sometimes experienced scientists who want to change research specialties with salary support, money for supplies, and travel expenses. At present, these grants are difficult to get (especially for experienced scientists) because of constraints caused by the budget deficit and the Iraq war. Despite this, such funds have directly resulted in the reduction of deaths caused by cancer.

Some of the earliest cancer research was done in the 1950s by George and Eva Klein from Sweden. These two scientists laid the groundwork for studies that define cancer (tumor) cells, as well as their recognition by the immune system. Their work continues to be groundbreaking and innovative (Klein and Vande Woude 2007). Other scientists through the years studied the factors that cause cancer.

Why Do We Get Cancer?

So what causes cancer? There are some simple answers to this question and some very complicated ones. Many aspects of this process called "tumorigenesis" involve the interruption of normal cellular functions by outside influences. Some of these influences come from our daily environment and some from biological sources. Let's start with the generalized sources and describe what they do to normal cells.

First, the elements from our environment. Sunlight, particularly in the form of ultraviolet light, is responsible for several forms of skin cancer (melanoma and basal cell carcinoma). Carcinogens such as asbestos, benzene, those in tobacco smoke, and many other chemicals can cause cancer: leukemia, lung cancer, colon cancer, and many other types. A type of environmental radiation called radon gas is the second leading cause of lung cancer. This radioactive gas is a by-product of the decay of certain types of uranium found in rock formations under buildings and homes.

The gas itself is very dense and can accumulate inside closed dwellings (Field et al. 2000). After discovery of the gas in the 1980s, preventive measures such as forced ventilation in houses have not only reduced the cancer risk but also contributed to a cleaner internal environment.

Most of these elements from our environment affect a cell's DNA. They cause breaks in the DNA and when they are repaired errors sometimes occur. These are called mutations. As these mutations slowly accumulate, they usually result in the production of mutant proteins within the cell and on the cell surface. Sometimes these changes are gradual and the cell remains in a pre-malignant or benign state for some time. However, when mutations occur in DNA that codes for growth regulation, it leads to unregulated cell division and growth. Such a cell is no longer a normal cell. It is a cancer cell.

Tumorigenesis can also be initiated by biological elements. Viruses can cause cells that they infect to become cancerous. Papilloma viruses can cause cervical cancer; hepatitis-C and hepatitis-B can cause liver cancer; human T cell leukemia virus can cause lymphoma. In addition, some viruses are associated with more than one type of cancer. The Epstein-Barr virus (EBV) is one of them. Most people have it and it has been shown to cause pharyngeal cancer, Hodgkin's disease, and a form of lymphoma. Viruses can cause a cell to become cancerous not only by inducing mutations, but also by blocking the pathway (inside the cell) of products produced by non-mutated DNA that are responsible for growth regulation. The result is the same and a cancer cell is created.

As mentioned above, rapid division by cells of the immune system is regulated. In some cases where hyper immune cell division happens (infections and organ transplantation), the same mutations in the DNA occur. This results in the development of lymphoma, a cancer often originating in mature T cells. The incidence of all of these malignancies is somewhat increased in people with weakened immune systems. Individuals infected with viruses such as human immunodeficiency virus (HIV), organ transplant patients, and the elderly may be unable to respond to an outbreak of cancer (Shultz 2009). However, cancers in these individuals can often be treated successfully despite their condition.

The Immune System and Cancer

The immune system is continually watching for threats. This threat can come in many forms: bacteria, viruses, other pathogens, and cancer cells. This watch is called "immune surveillance" (Burnet et al. 1971), the continual search for an entity called non-self. Normal cells in the body are recognized as "self" during this process. When these cells become cancerous they produce proteins that the immune system cannot recognize as self. The surveillance system, consisting of three types of immune cells, begins to take notice. They are the natural killer (NK) cell, the T cell, and the antigen-presenting cell (APC). The APC is usually called a macrophage, and there are many variations of this cell type. The NK cell can recognize an irregular (for lack of a better word) cell surface associated with the tumor cell. It then promptly bores a hole in the tumor cell, killing it. The effect is rather indiscriminate and short-lived, but the initial death of some tumor cells is very important. The APC ingests the dead tumor cells, finds the non-self proteins, and processes them into peptides (10–20 amino acids long).

The T cell is the primary enforcer of the cellular immune system. It can kill its target in three different ways, regulate other T cells, control APC function, and actually change the body's temperature to make viruses uncomfortable. A T cell cannot respond to non-self unless it sees non-self associated with self. The APC links non-self peptides to self-proteins and they get put on the APC surface where T cells can see them. This results in the activation of the T cells.

In addition, some normal cells have the ability to alert the immune system that they are becoming cancerous. As a normal cell begins to turn into a cancerous cell, it begins to act as an APC in an apparent attempt to save the body. It does something it never does as a normal cell: express self-proteins and the tumor's non-self proteins on its cell surface. Once a T cell sees this combination of non-self/self-protein, it begins to mature into a killer T cell capable of killing any cell (cancer) that has that non-self peptide on its surface (Alexander et al. 1989).

T cells mature into killer T cells thousands of times over our lifetime, but this is difficult to detect because it involves so few cells. Once the cancer cells are destroyed, the T cell does one more remarkable thing: it matures even further into a cell that retains the memory of the cancer and has the ability to live for years. If the cancer returns in, say, five years, the memory T cell awakens. It then recruits the rest of the immune system to quickly remove the cancer cells from the body, faster than the first time. However, the tumor has many ways to subvert the immune system. It is important to understand and know what these subversive tactics are in order to design a defense against them.

How Cancer Cells Subvert the Anti-cancer T cell Response

The process that turns a normal cell into a cancer cell occurs in stages until unregulated growth begins. Sometimes, mutant proteins are created that are similar to self-proteins, but not quite self. The immune system gets confused. It begins to respond, but it's actually responding to self. For instance, when some melanoma patients respond to their tumor, normal pigment cells (melanocytes) are destroyed, resulting in vitiligo, the formation of white patches. The immune system then has to correct this mistake. In the meantime, the tumor has diverted the attention of the immune system and multiplies. There is more on this subject in Chapter 13, where the role of the cancer itself in cancer treatment is examined.

During this diversion, the tumor cell often becomes more lethal by leaving the primary site for another distant location within the body (metastasizing). As the immune system applies pressure on the tumor, this group of cancer cells changes and evolves. The immune system can actually select highly lethal tumor variants that are invisible to the immune system. This process is called "immune selection" (Loose and Van de Wiele 2009). It occurs because the cancer cells behave like stem cells (see Chapter 8 on cancer stem cells). The stem cell nature of the cancer allows it to produce subpopulations of cells that are distinctly different from the original cancer cell. Once the immune system applies pressure on one type of cancer cell, another slightly different cancer cell appears. The slight differences confuse the immune system, which allows the cancer to grow. The immune system tries to keep up, but sometimes it can't. Immune cells can travel to almost any location in the body and seek out and try to destroy the cancer cells. Sometimes they are at a distinct disadvantage because they can no longer see the tumor cells. Some methods are now being developed to allow the immune system to function under these conditions (Frank et al. 2010).

The Development of Chemotherapy

Early treatments for cancer used chemical agents (chemotherapy) to retard the rapid growth of cancer cells. The chemical treatments were developed by scientists who identified the fact that cancer cells have a higher metabolism than normal cells. Cancer cells take in more nutrients and substances at a higher rate than their normal counterparts. Scientists found that if they administered low doses of a toxic chemical, the tumor cell would absorb more of it and die. But prolonged exposure to the chemical agents produced harmful side effects.

One agent used in early treatments was called mitomycin-C. It was a by-product of a certain type of bacterium and later made in the lab. It has the ability to cross-link the DNA chains. The cell is unable to divide because the separation of DNA is an essential component of cell division (mitosis).

In early studies, the doses were too high and sometimes led to the death of the patient. When the chemical is dissolved it displays a deep blue color. In the initial trials, it was clandestinely termed "The Blue Death" after a rat poison. Extensive clinical trials have proven that at the proper dose and rate it is effective against esophageal cancer (Enzinger et al. 1999), breast cancer (Vogl et al. 2010), and bladder tumors (Bohle 2010). It also is used in the laboratory so that scientists (mostly tumor immunologists) can study immune T cell responses to tumor cells (Chanarin 1989). Since the tumor cells cannot divide after mitomycin-C treatment, the only measurable cell division comes from the stimulation of T cell nuclear replication and cell division induced by the inactivated tumor cells.

The evolution of mitomycin-C from toxic chemical to a useful chemotherapeutic agent took years. Today, chemical agents used for cancer therapy are designed to attack specific tumor cell products that are unique to the tumor, which can lead to the death of only the cancer cell. This type of chemical therapy reduces the side effects normally associated with use of broad-spectrum chemical agents. Broad-spectrum agents would often weaken or severely deplete the very cells (immune cells) in the body capable of removing the cancer cells. Broad-spectrum chemotherapeutic agents, including the ones mentioned above, are toxic to the body to one degree or another. The problem is the age of the patient. Two patients with the same cancer diagnosis but 35 years apart in age may get different modes of treatment or different doses of the same chemical. The younger patient (say, 20 years old) can withstand a higher dose for a longer period of time than the older patient (55 years old). Therefore, the prognosis (possible outcome) of the same type of cancer can be better for a younger person. This is simple physiology, but there are exceptions and in some cases completely opposite results. This is the nature of chemotherapeutic treatment. Scientists are now recognizing that the response of a tumor to chemotherapy is related to the genetic make-up of the patient. Scientists noticed that patients with the same type of cancer who responded positively to a given agent had identical genetic markers. So, in a new approach, they could design these drugs around the genetic make-up of the patient (Glynn et al. 2009). This is an ever-changing field of scientific research, which will be joined by a newer type of cancer treatment called immunotherapy.

The Advent of Immunotherapy

What is immune-based therapy (immunotherapy) and how did it start? It began in mid-1970 when scientists were studying a mouse tumor model. The tumor was blood-borne and usually lethal, but they found an active immune response to it. The scientists grew the tumor cells in the lab, inactivated them using mitomycin-C, and cultured them with normal mouse lymphocytes (T cells) over several days in a temperature- and gas-controlled incubator. They found that the T cells had become immune to the tumor and had the ability to kill them.

Subsequently, massive amounts of these T cells were produced. Mice were inoculated with a small number of live tumor cells and injected with tumor-immune T cells. Mice that received the immune T cells survived much longer than those that did not. The technique was called "adoptive transfer" (June 2007) and it ushered in a new approach to treating cancer. Recent advances in genetic transfer have allowed scientists to genetically engineer T cells to make them recognize the tumor. This means that normal T cells can be removed from the body, genetically altered, and then returned to the circulation to fight the tumor (Zhao et al. 2005). The technique is used in conjunction with natural immune-stimulating proteins such as lymphokines. Natural processes of the immune system can thus be manipulated to fight previously lethal cancers. As we move forward, this concept will get a bit more complicated because of the inter-relationships between the immune cells that make immunotherapy effective at removing cancer from the body.

Immunotherapy of Human Cancer

Immune-based therapy for cancer in humans has been going on for 20 years. Despite this, the procedures are still in the early stages of development. Immunotherapy for cancer is just beginning to be recognized as the future of modern cancer treatment. Immunotherapeutic protocols are conducted in phases. A Phase I trial is used to determine how the treatment is given and what dosages (amounts) should be used. These trials relate to the safety of the treatments and the participants are those who have not been cured by standard treatments for the kind of cancer they have. Phase II trials include the same types of patients, but more attention is paid to the effect that the safe (Phase I) protocol has on the growth of the cancer itself. Phase III trials compare the effectiveness of the new therapy with the standard treatment protocols. These treatments are random blind studies, where patients and sometimes physicians do not know what treatment is being administered. If the new treatment is found to be superior to the standard protocols early in the trials, the study is halted and the standard treatment group is given the new treatment protocol. Phase IV studies often involve patients who are not as far along in their disease as those in Phase I and II clinical trials. Phase IV studies actually refine the techniques used in a successful Phase III trial and are designed to maximize the anti-cancer effect of the treatment (Turkington and LiPera 2005).

One aspect of an immunotherapeutic clinical trial is the human factor. The trials are now conducted with patients who have late-stage cancer. The early trials (Phase I and II) are more concerned with the safety of the procedure than the effect on the cancer. Remember these are experimental procedures that have never been performed with human subjects, yet these patients volunteer to be in the study. The patients study and sign an informed consent agreement, which informs them of the possible side effects of the treatment. The volunteers are willing to help others benefit from the study that may not help them survive. This altruism is at the heart of modern clinical trials. None of the new immune-based therapies can be used unless their safety is stringently verified by early trials. The participants are as important in the development of a treatment as the scientists and physicians who develop and administer it. They are, in a sense, pioneers exploring this new type of cancer treatment called immunotherapy.

Scientifically, immunotherapeutic clinical trials are broken down into three areas. Each involves the recruitment of immune cells to fight cancer by either augmenting natural immunity or creating a situation where immunity to cancer is artificially induced. The techniques are sometimes complex, but their effect is simple. The goal is to invoke the two most powerful weapons the immune system has to fight cancer. Both weapons involve the T cell, but its action is guided by the manipulation of other cells of the immune system that prepare the way. One of the weapons is the killer T cell. Once it identifies a cancer cell peptide, it goes about the job of killing the cancer cells that express the peptide. The second weapon is the biological memory of the cancer embedded in the DNA of the memory CD4 and CD8 T cells. This memory is powerful and results in the rapid destruction of the recurring cancer. The memory can also last a lifetime. The three areas mentioned above involve the manipulation of the APC or macrophage, the T cell, and the cancer cell itself. All these areas individually hold promise as components of an integrated immunotherapeutic treatment.

The cellular components are manipulated in several ways. The APC, because of its ability to activate anti-cancer T cells and induce killer T cells, can be removed from the body and treated in the laboratory. The first treatment turns it into an efficient APC (called dendritic cell or DC) using a bioactive protein called a lymphokine (granulocyte-monocyte colony forming factor). The DCs are 20 times as efficient at presenting cancer antigen peptides. During the second treatment, they are fed the patient's cancer cell membranes and rapidly begin to present the cancer peptides. The cells are returned to the patient along with another lymphokine that activates killer T cells (Robson et al. 2010). Phase I and II clinical trials have proven that this procedure is relatively safe as long as the dose of the lymphokine is kept low. If not, patients often develop a fever and other minor symptoms. Significant immunity to the cancer has been demonstrated using this type of immunotherapy.

Direct manipulation of the T cell is underway in a number of clinical trials where the T cells are removed from the cancer patient and genetically altered to recognize the patient's cancer peptides. Separate clinical trials have also been performed using these T cells that have a second gene, which produces the killer T cell lymphokine (called IL-2) mentioned above. After the T cells have been tested to determine whether they are producing the proteins from the gene transfer, they are given back to the patient. This procedure (protocol), which has been used in the same Phase I and II clinical trials as above, has been shown to induce moderate decreases in the size of the cancer (Morgan et al. 2006). The technique of genetically altering T cells is complex and needs to be refined. There are problems with the T cells retaining the genetically engineered information after their encounter with the cancer. As soon as this encounter occurs, the T cells begin a normal process of rapid proliferation. Often this will cause the inserted genes to get thrown out of the cell because of the highly efficient cell division process T cells go through. Once the genes are lost, the anti-cancer effect is minimized to the point where it is ineffective. These problems as well as others associated with this protocol are being addressed with the hope that one day this technique will result in significant reduction of the patient's cancer cell numbers (deWitte et al. 2008).

Manipulation of the cancer cells themselves in immunotherapeutic trials also has shown promise. Without much effort, some cancer cells can be removed from the patient's body and genetically altered. In this case, the cancer cells are given several different genetic modifications. These modifications not only cause the T cell immune system to recognize them, but also stimulate anti-cancer T cells to increase their numbers. One of the proteins expressed by the inserted genes is a self-protein needed by anti-cancer T cells to respond to the cancer. Often there is no expression of these self-proteins on the surface of late-stage cancers. The other protein secreted is one or several of those small bioactive proteins called lymphokines. These genetically altered cancer cells are returned to the patient's body. This time, they have been rendered unable to divide by the use of irradiation or chemicals such as mitomycin-C. This combination of proteins made by the altered cancer cells is supposed to drive the patient's T cell immune system to attack the growing cancer (Culver 1996).

These three areas of immunotherapeutic manipulation of immune cells are by no means the extent of this effort to redirect the T cell immune system to eliminate the cancer. Efforts are underway to temporarily inhibit anti-autoimmune T regulatory cells, which can inhibit anti-cancer T cells from attacking the cancer (Wolchok and Saenger 2008) by the use of anti-cancer T cells created from stem cells (Cheong et al. 2002), and the activation of "death receptors" expressed by the cancer cell (Screpanti et al. 2005). The hope is that one day these manipulations of the immune system will be combined to completely destroy the cancer and have it never return because of the memory left behind.

Chemotherapy and Immunotherapy Combined

It has been shown that some of the new chemotherapeutic agents that kill tumor cells can render the dead tumor cells immunogenic. So as the tumor dies the immune system is activated by what is now a vaccine. This leads to the destruction of even more cancer cells. This is the newest variation of the use of cancer vaccines and chemotherapy to treat advanced cancer. The most important aspect of this type of therapy is the production of memory T cells, which will guard against further outbreaks of the tumor. To further examine the future use of immune-based therapy and chemotherapy, it is important to look at another type of tumor that circulates in the blood.

There are many blood-borne cancers. Leukemia (see Chapter 6, Immunotherapeutic Approaches to the Treatment of Leukemia) is one of these cancers. It originates from the bone marrow, which is the birthplace of both white and red blood cells. Unfortunately, the people who most often get leukemia are young children and the elderly. Some forms of leukemia are curable, especially in young patients. The treatments can be painful and have many side effects. At first, the treatment for leukemia makes use of chemotherapy. If that fails, a bone marrow transplant is the next step. Since the leukemia cells originate in the bone marrow, it is necessary to remove all the cells in it and replace the marrow itself. The donors of the bone marrow are often, but not always, matched to the genetic type of the recipient. The difficult part is next. The patient's bone marrow must be completely destroyed prior to the transfer of the donor marrow.

This treatment involves a combination of chemotherapy and radiation, which leaves the patient devoid of white blood cells and susceptible to life-threatening infections. The donor marrow is infused into the patient and the cells begin to reside in the marrow of the patient. The donated bone marrow cells have the ability to attack and remove any residual leukemia cells because the donor's T cells recognize them as foreign. However, even if the donor is an identical genetic match to the patient, there is a possibility that the donated marrow cells will begin to attack the patient's tissues. This is because very slight genetic changes occur to our cells after we are born. This condition is called graft vs. host disease and can be deadly. At present, the patient must be given anti-rejection (immune suppressing) drugs until the grafted cells establish themselves (Barton-Burke et al. 2008).

Another instance where the two therapies are used together has to do with the previously mentioned "death receptors". Normally the death receptor is expressed by a wide variety of normal cells. It is used by the immune system to cause rapid death of cells infected with viruses. This type of cell death results in the degradation of the target cell's DNA. That is because some viruses have the ability to integrate their viral DNA with the infected cell's DNA. This is a way to not only prevent the virus from reproducing and spreading to other cells, but also remove viral DNA, which, if left intact, could infect any cell (macrophage) that would ingest the dead cell. This ligand (key) that attaches to the receptor (lock) is expressed by T cells and can be secreted. Unfortunately, many cancer cells do not express the receptor. However, a chemical agent called Topotecan or TPT can induce the expression of the receptor on cancer cells. The cancer cell treated with this agent dies and its proteins become available to the immune system. Immunotherapeutic trials have shown that TPT is safe to use and can result in the reduction of the cancer size in many patients. Chemotherapeutic agents like TPT are being developed because generally they have fewer side effects and induce the body's own defenses to fight the cancer (Ackerman et al. 2007).

Today, because of the advances made in laboratory technology, it takes only months to develop agents like TPT. This is an example of a chemical agent used in anti-cancer T cell therapy and designed to alter specific tumor cell products that lead to the death of the cell. This type of chemical therapy reduces the side effects normally associated with the use of broad-spectrum chemical agents because the cancer cells can be treated in the lab and inactivated cells returned to the patient. Even though TPT has been shown to be safe to use like traditional chemotherapeutic agents it is an advance in anti-cancer therapy. It is possible to culture cancer cells in the lab for use in immunotherapeutic anti-cancer treatments. In these cases, both immunotherapy and chemotherapy can be used to help eliminate the cancer and also to improve the quality of life for patients who undergo treatment.

The Future of Immunotherapy

In the future, cancer treatment should be much less painful and free of many of the side effects patients have had to endure. New technologies are continually being developed to deal with this dreadful disease. Eradication of some types of cancer is a reality today, but a few years ago the diagnosis was a death sentence. More research must be done in order to eradicate cancer. The questions are how much research and how long will it take? Scientists seem optimistic about the future of cancer treatment, but they also realize that it will take a major commitment by the government and the private sector. Many of the issues involved in cancer treatment have less to do with cancer than with finances. Some years ago, government policies prevented thousands of cancer patients from participating in possibly life-saving clinical trials just to save money that eventually would be spent on war. This is just a small example of the forces at work as far as cancer research is concerned. The private sector is involved as well.

There must be a balance between government support of cancer research and the profit (of pharmaceutical companies) associated with it. Government-supported research works toward a more theoretical type of cancer research and pharmaceutical research toward its practical application. Discovery and innovation lead to application. Without them, there will be no new ideas to turn into life-saving drugs. Over the past few years, government support for cancer research (in the form of grants and new clinical trials) has slowed dramatically. During that time, the cancer rate was still high.

However, there is light at the end of this tunnel. The Obama administration (Democrates 2009) have earmarked billions of dollars toward continued cancer research. This money, it is hoped, will go towards re-establishing the National Cancer Institute's Clinical Trials Cooperative Group's program. Years ago, this program was ordered by the George Bush Jr. administration in 2003 to halt 60% of its proposed new clinical trials. As a result, 2,000 to 3,000 cancer patients were left out of new clinical trials. Hundreds of them might have been alive today if they had been given the chance to participate. Because of the time it takes to develop new anti-cancer treatments, the new therapies that would have been studied during that period could have by this time been used to save lives.

The development of new immunotherapeutic treatments for cancer and their use in clinical trials are essential. As we have said, 2,000 to 3,000 people die each day from cancer in the United States. Over the next 20 years that number will increase by a factor of 10 in five-year increments because by that time 25% of our population will be over 65 years of age. The older you get, the more likely it is that you will get cancer. These new immune-based therapies have been proven to be not only less painful, but also less deadly to the patient. Some elderly patients are denied chemotherapy because the treatments are so toxic that those patients may actually die from the treatment instead of the cancer. This is one of the reasons why immunotherapy of cancer should be one of the most important areas of research for cancer scientists and clinicians. Without it, many of us may face a painful cancer death.

Cancer has an impact on our entire society, not just those who contract the disease. Think for a minute of those people (some of them famous) who have died of cancer. They made significant contributions to all of humanity. Many of them died at the peak of their brilliance and at a young age. Their contributions to the world just stopped, and all of us suffer for it. Cancer does not discriminate according to race, social status, and wealth. It is an equal opportunity killer.

Why was funding for cancer research cut? Why is some research associated with cancer banned? Why, knowing that you could get this disease and die, would you vote for or condone funding reductions for cancer research that some day may save your life? These are good questions, obviously asked by someone who does not understand. This is the society we live in, but the fortunate thing is that it can change and probably will.

The other aspect of cancer death is a personal one. Families and friends have been devastated by the loss of a loved one in ways that can hardly be described. Friendships are lost and vibrant lives are taken away from all of us by this disease. Sometimes, the people left behind are the unlucky ones because they have to live with the memory of a loved one suffering from cancer. However, those lost to this disease would most certainly want us to get on with our lives and remember them as they were, a loved one, a friend. Those of us who survive owe it to those lost to do everything possible to bring an end to cancer. That goes doubly for those who are both survivors and cancer researchers, like me.

Can immune-based therapies save us from cancer? The answer is probably yes. However, this is dependent on the ability of theoretical cancer research scientists to harness the power of the immune system to defend the body from this disease. If we can do this, in the future when we hear that a friend or family member has been diagnosed with cancer we can feel confident that they will be treated successfully and will be able to return to their normal lives.

Chapter 6

Immunotherapeutic Approaches to the Treatment of Leukemia

Leukemia is a malignant proliferation of blood leukocytes, usually characterized by leukocytosis and infiltration of other organs by leukemic cells, ultimately causing death. This is the standard dictionary definition of leukemia, but leukemia is much more. Its impact on our society has been formidable. There are magazines and scientific journals just about leukemia. There are also societies, patient groups, and survivor reunions, all associated with leukemia. Leukemia is one of many types of cancers, but it has permeated our society in a way that sets it apart from all other forms of cancer. There are many reasons for this. For one thing, it is a very difficult type of cancer to treat because of its nature. Another reason is that, during treatment, there is greater anxiety experienced by both patient and doctor than during other cancer treatments. Treatment takes longer and patients and their families go through constant remission/relapse cycles. More adults are stricken by leukemia, but it accounts for only 30 percent of childhood cancers. However, when children suffer we suffer (Westcott 1999).

Technically, there are two stages to leukemia. The first is the chronic stage, where abnormal white blood cells (leukocytes) can still function and there is a slow onset of the disease. Many people who are in this stage do not know it because there are no symptoms. The next stage is acute leukemia, where the leukocytes cannot function normally and their numbers rapidly increase. This stage of leukemia worsens quickly with the onset of symptoms such as fatigue due to the lack of red blood cells, easy bruising due to low numbers of platelets (which control bleeding), and the recurrence of minor infections because of low numbers of functioning white blood cells (lymphocytes).

Leukemia affects white blood cells in the bone marrow, where they are born. The two major types of white blood cells (leukocytes) in the bone marrow are the lymphoid (lymphocyte) and the myeloid (myelocyte) types. There is acute and chronic lymphocytic leukemia and acute and chronic myelogenous leukemia. There are about 200,000 people living with leukemia and in 2008 there were an estimated 50,000 newly diagnosed cases in the United States. Leukemia is actually diagnosed more in adults than in children, at a ratio of 10:1. There were about 20,000 deaths from leukemia in 2008. The five-year survival rate (cure rate) today is about 50%, up from 15% in the 1960s. This is due to dramatically improved treatments for this type of cancer. What causes leukemia is not completely understood. It is known that job-related exposure to benzene and exposure to ionizing radiation can cause leukemia, but this does not explain the vast number of cases (Turkington and LiPera 2005). What is important is that the cure rate is improving. A contributing factor to the increase in cure rate is the development of new therapeutic treatments.

Immunotherapeutic approaches to the treatment of leukemia are a recent development. They are an offshoot of the adoptive transfer experiments of the 1970s. T cell-based immunotherapy for leukemia presents special problems because of the lymphocytic origin of this disease. In this chapter we initially focus on the nature of the T cell immune response to leukemia while describing the problems the T cell immune system faces in fighting this type of cancer. We also examine current immunotherapeutic treatments for leukemia and what role they play in the overall leukemia treatment methodologies. We then look at the newest experimental approaches to T cell-based therapies and determine what advantage they have over current treatments for leukemia. Finally, we ask what is the future of T cell-based therapies and what possible impact they could have on the management of leukemia.

The T cell Response to Leukemia and Problems Associated with that Response

Leukemia presents unusual problems for the T cell immune system. First of all, the leukemia, being derived from the lymphoid lineage of bone marrow leukocytes, appear as self to the responding T cells. Even if the T cell immune system did respond to these cells, the anti-autoimmune arm of the T cell immune system (Tregs) would shut them down. This is a serious problem for the T cell immune system. One tiny advantage is that this type of cancer, like many others, has a "stem cell quality" about it. That means the malignant cells tend to produce altered proteins (peptides) that are quite unnatural and can be seen as non-self. Once this occurs, the T cell immune system begins to attack such cells, but the cells that produce the altered peptides are in a minority. They are quickly killed, leaving scores of untouched leukemia cells, which continue to grow (Frank et al. 2010).

As these cells grow, they produce another population with another altered peptide and they are killed as well. This becomes a constant cycle in which leukemia populations are killed while other populations continue to produce more recognizable targets. This begins to exhaust the T cell immune system to the point where T cell numbers begin to dwindle, and such a case is referred to as "immunological exhaustion". This condition can contribute to the recurrence of minor infections because the number of T cells is reduced (Aisenberg 2006).

Myeloid leukemia evokes a different response from the T cell immune system. Because the cells are not lymphocyte-like, the T cell immune system can see them. However, it is only a limited vision. Myeloid leukemia cells begin to proliferate in the stem cell fashion described earlier, but they seem to display more peptides that the T cell immune system can recognize as non-self. In this case, the T cell immune system has a foothold, something it can see and respond to. It turns out that the myeloid leukemia begins to express a peptide associated with an enzyme that degrades other proteins called proteinase 3. Of course, this is a part of the stem cell program (Ghosh and Matsui 2009).

Normal myeloid cells do not express this peptide, but because of the cancerous nature of the leukemia, this expression is forthcoming. There is a caveat to this scenario. The T cell immune system generates two types of killer cells for this particular leukemic peptide. One has a high affinity for the leukemia and can efficiently kill it. The other one has a low affinity. Myeloid leukemia can eliminate the high-affinity T cell by expressing an overabundance of the peptide on the surface. This is actually a normal mechanism in the body because it is used to control the T cell immune response. However, in this case the myeloid leukemia uses it to subvert the T cell immune system (Dao et al. 2008).

Current Immunological Treatments for Leukemia

Some of the current immunotherapy concerning leukemia has focused on graft vs. host disease (GVHD). This disease occurs when bone marrow has been transplanted to treat leukemia (James and Odom 1983). Patients who have this disease described to them by their doctor have a hard time picturing what could happen to them during the transplant. Since the leukemia cells originate in the bone marrow, it is necessary to remove all of the cells in it and replace the marrow itself. The donors of the bone marrow are often, but not always, matched to the genetic type of the recipient. The difficult part is next.

The bone marrow of the patient must be completely destroyed prior to the transfer of the donor marrow. This treatment involves a combination of chemotherapy and radiation, which leaves the patient devoid of white blood cells and susceptible to life-threatening infections. The donor marrow is infused into the patient and the cells begin to reside in the marrow of the patient. The transplanted donor bone marrow T cells recognize the leukemia and kill it. Again, this is because of two things. First, there are small genetic differences even between two matched donors. Second, the donor's T cells have never seen the leukemic cells before. There is one more thing to be concerned about. Even if the donor is an identical genetic match to the patient, there is a possibility that the donated marrow cells will begin to attack the patient's tissues. This is the definition of GVHD. Research is being done to identify the T cells that mediate this process and remove them prior to the transplant.

These T cells see very slight genetic differences between the donor and the recipient even if the two individuals are twins. The differences are due to very minor changes that occur in our DNA after we are born. The donor T cells see the protein products of the recipient's DNA as non-self. This triggers a response and tissues with these proteins are attacked and damaged. Those tissues may be vital organs such as the intestine, liver, or skin. Currently, anti-rejection (immune-suppressing) drugs are used to control GVHD, but they have undesirable side effects, such as susceptibility to minor infections (Turkington and LiPera 2005). Once scientists identify the protein targets of the tissues under attack, it will be simple to remove those T cells from the donor's marrow. This will reduce or eliminate the life-threatening effects of GVHD and allow the donor's anti-leukemic T killer cells to attack and eliminate the cancer. Refinements to techniques like this one will minimize or eliminate the use of anti-rejection drugs and the side effects associated with them.

The Newest Approaches to T cell-based Immunotherapy for Leukemia and Their Advantages over Existing Treatments

Some new immunotherapies for leukemia are already being used. They involve the use of cellular vaccines. This technique uses the bone marrow of the leukemia patient. The bone marrow produces the leukemic cells. The cells are either of lymphoid or myeloid origin. However, there is another cellular lineage that is not affected. This is the granulocyte-monocyte lineage. These cells turn into macrophages. They have the ability to go anywhere in the body. Macrophages can even go where some T cells cannot. For instance, they can freely travel in and out of the brain through the blood-brain barrier. This cell plays a pivotal role in the induction of anti-leukemic T killer cells. Ideally, the macrophage would eat the leukemia cells and present the non-self proteins (peptide subunits) on its cell surface. Once the T cells see this they begin to generate a response that results in the production of T killer cells, which will eliminate the leukemia.

Scientists have manipulated this process using a protein expressed by leukemia cells. This protein is called Wilm's Tumor Antigen-1 (WT-1). Once the macrophages are isolated from the patient, the gene for the leukemia-associated protein is inserted. The macrophages are then put back in the patient and allowed to express the protein, at which time the T cells begin to respond and kill the leukemic cells. The WT-1 protein is actually a kidney protein. Leukemia cells, because of their stem cell nature, express this protein (Dao et al. 2008).

Another type of immunotherapy for leukemia (myeloid) is being used on patients in stage 2 clinical trials. This type of immunotherapy makes use of the patient's own leukemia cells. A single leukemia cell is isolated and its numbers multiplied in the laboratory. Once enough cells are grown, a gene is inserted into them. The protein produced by this gene is an immune-stimulating growth factor called GM-CSF. The protein-producing leukemia cells have their growth arrested, but they can still produce protein. With cell division inactivated, the leukemia cells are re-introduced into the patient. Results of the technique have been favorable. This type of immunotherapy is primarily used after successful standard chemotherapy and has been shown to prevent relapses in patients who, without this treatment, would suffer a recurrence of the disease (Wierda 2007).

The potential advantage these therapies have over conventional chemotherapies is obvious. They can either replace the chemotherapy or minimize its use. The list of side effects from the various chemotherapeutic treatments is extensive. They can be temporarily disfiguring, painful, and sometimes lethal in older patients. Therefore, the logical option is to use a therapy that does not cause these types of side effects. The other and very important advantage of T cell immune-based therapy is the creation of a T cell memory after the cancer is eliminated. This T cell memory system is rapid and at least 10 times as effective as the primary response in removing a recurring cancer. So it is apparent that once immunotherapeutic approaches to the treatment of leukemia begin to supplement or replace standard chemotherapeutic treatments, the incidence of painful and potentially lethal side effects should be reduced.

The Future of T cell-based Immunotherapy for Treatment of Leukemia and Its Impact on Leukemia Management

T cell-based immunotherapy will play a major role in the treatment of leukemia. Newly developed immunotherapies for leukemia will increase the cure rate for this disease to well above 90 percent, or even 100 percent. It may be that once you have been diagnosed with leukemia you will go through a couple of painless treatments and then get on with your life. It has happened with other diseases.

In 1988, many people were dying from human immunodeficiency virus (HIV). During and after this time, pharmaceutical companies developed many drugs to fight the virus. Almost none of the drugs worked alone as hoped, but a combination of the drugs could completely stop the virus from replicating. Now, you do not die from HIV, you live with it. This can happen with leukemia, but in another way. Cancer, particularly leukemia, is different. Cancer is derived from our body and not contracted, as with a virus. It is possible that the only way to ultimately cure or treat leukemia is by manipulating the body's T cell immune system. Leukemia and many other cancers have a death grip on our society. Almost 3,000 people in the United States die each day from cancer and almost 4 percent of them from leukemia (~55,000 a year).

Someone you know may get leukemia. The first thing you will do is panic. The problem will be getting that person to the right hospital to see the right doctor to get the best treatment. Treatment is a complicated process requiring a great deal of patience. Remember, the doctors and the hospital have the patient's best interest at heart, though it may not seem like it. The family and the patient must be well informed, because sometimes there are difficult decisions to be made. Without this knowledge the tendency is to give up. That is not good. Where there is life there is hope. With the advent of new immunotherapeutic approaches to the treatment of leukemia, it may be easier to sustain this hope. Our entire society will benefit from the reduction of suffering of cancer patients.

Chapter 7

Cancer-induced Autoimmunity

Autoimmunity is a disorder of the body's defense mechanism in which an immune response is elicited against its own tissues. This is a definition from the *Oxford Dictionary of Biology*. At some point in your life, you will probably experience some problems related to autoimmunity. Arthritis is an example. The pain associated with this disease can be severe. If you develop cancer, there is a high likelihood you will suffer from some type of autoimmune reaction. When a cancer cell develops, the T cell immune system has a problem. It will ask itself, "Is this cell normal or abnormal?" This may take a long time. Sometimes, by the time the T cell immune system realizes that the cell is cancer, it is too late. If not, the T cells move in to attack the cancer, but the system is still quite confused. That confusion causes normal cells to be destroyed. This is the short definition of cancer-induced autoimmunity.

For some reason, one particular type of cancer derived from the cells of the neural crest causes autoimmunity at a higher frequency than other cancers. That cancer is malignant melanoma. Melanoma cancers originate from normal pigment-producing cells called melanocytes. During the T cell immune response to melanoma, some of the T cells attack normal melanocytes because of their similarity to the melanoma cancer. They are destroyed, resulting in vitiligo or loss of skin pigmentation (Fujimoto et al., 2006). Other cells closely related to melanocytes, such as cells of the nervous system, are also attacked, causing the painful condition called peripheral neuropathy (Visovsky et al. 2005). During these episodes, the immune system's attention is focused not only on these two responses, but also on how to stop them through the recruitment of regulatory T cells that are designed to shut down such responses. This confusion allows the cancer cells to multiply.

It is apparent from this example that anti-cancer T cells can cause problems while they are trying to kill the cancer. However, another part of the T cell immune system sees these attacks as anti-self and tries to stop them. Sometimes all the activity of anti-cancer T cell response is suspended because of this fact. In this chapter we discuss what happens when the anti-cancer T cell response is suspended. The immune system's control over itself is at the core of cancer-induced autoimmunity. If the immune system can properly control the actions of anti-cancer T cells, there is no autoimmunity and the cancer is eliminated. Both autoimmunity and cancer can be deadly, but often you cannot have one without the other because of the complex nature of the T cell immune system. We will outline just what autoimmunity is and then examine what efforts are made to control or cure it. Also, we ask, why does getting cancer involve autoimmunity, and how does the T cell immune system try to prevent it? Finally, what are the consequences of preventing anti-cancer T cells from inducing autoimmune reactions?

What Is Autoimmunity?

Autoimmunity is the result of our immune system seeing self as non-self. There are many immune cells that participate in this response. The first cell, which is pivotal in the response, is the macrophage. It is called an antigen-presenting cell (APC) because it actively scavenges the body and has the ability to reduce bacteria, viruses, and cancer cells into protein components called peptides. These peptides are what trigger the immune system to react. The peptides have to be expressed on the surface of the APC linked to a self-protein called HLA-DR (peptide-self complex). The T cell, in this case the CD4 T cell, detects this peptide-self complex. It responds to the complex by becoming activated. This activation step precedes cell division but includes the secretion of bioactive proteins called lymphokines.

One of the lymphokines is called IL-2 and it is needed by CD8 T cells to effectively kill cells expressing a similar peptide-self complex, except that the peptide is attached to the self-protein HLA-ABC. The CD-8 T cell kills virus-infected cells before they release other viruses. They also kill cancer cells. Both of these target cells must express the peptide-self complex, which is seen as non-self by the CD8 T cell. The last cell is the B cell, which makes antibodies in response to other bioactive proteins secreted by the activated CD4 T cell. The antibodies can attach themselves to bacterial and cancer cells. The macrophage can then ingest these cells and start the process of antigen presentation. Our discussion centers on the T cell immune system because it is largely responsible for the entire immune response.

The T cells function to protect us from bacteria, viruses, and cancer. However, there are cases in which the peptides derived from each entity are so similar to our self-peptides that the T cells begin to attack normal tissues, an autoimmune reaction (Gorski et al. 2001). Even though there are other ways in which autoimmunity is induced, which will be discussed later, this is a major method of induction.

A classic example of autoimmunity was the flu pandemic in the 1900s, which resulted in the deaths of millions of people. Peter Doherty and Rolf Zingernagel of Australia won the Nobel Prize for their discovery of the HLA proteins. They described how killer CD-8 T cells attack and kill flu virus–infected cells. They also determined that the millions of deaths during the pandemic were not actually caused by the virus itself, but were the result of T cells attacking normal cells. The virus induced such a strong immune response coupled with the exposure to human peptides that the T cells lost the ability to distinguish between self and non-self. Therefore, influenza-activated T cells attacked normal tissues, which resulted in many deaths. This type of autoimmune reaction is very serious. It is feared that one day such an epidemic will happen again. Scientists are carefully watching the "bird flu". They are hoping it will not recombine with human DNA and cause another pandemic.

Even though this is not an example of cancer-related autoimmunity, it shows how the T cell immune system can be harmful. We survive because our immune systems are efficient. This efficiency can cause some problems to our health. There is a major effort to control our T cell immune system by using the immune system itself, which will be discussed later. However, medical efforts to manage autoimmunity started out using a natural protein.

Efforts to Control or Cure Autoimmunity

Autoimmunity has been a problem for a very long time. It was initially thought that a natural cure could be obtained in the form of steroids. Steroids can reduce inflammation, which is often a starting point for autoimmune reactions, but they presented problems that compromised their use for anything other than inflammation. It was found that these proteins had to be used only periodically or they could cause severe reactions. Steroids can destroy many normal tissues. They can actually induce a form of autoimmunity that is deadly. Steroids can induce diabetes because they cause the cells that produce insulin (pancreatic cells) to become abnormal. This abnormality is recognized by the T cell immune system as foreign, and the cells are destroyed, leading to a lack of insulin (Iwamoto et al. 2004). Many people who receive steroids for lung inflammation (COPD or pre-emphysema) inadvertently get diabetes as a result of the treatment. It is true that steroids can reduce inflammation, but there are dangers. The root of this problem has to do with the T cell immune system. The problem of cancer-induced autoimmunity is the same problem that causes lupus, arthritis, diabetes, and many other autoimmune disorders.

The solution may not have to be expensive or complicated. We know that the T cell immune system is supposed to police itself. We do not know why T cells (Tregs) that are supposed to stop autoimmune reactions cannot. Our immune system is designed to work like many other biological systems. Some of this occurs as we get older. Some of the T cells controlling autoimmune T cells go missing. We need a way to actually rejuvenate our immune systems.

Heath foods and supplements that are supposed to increase the activity of the T cell immune system are currently not approved by the Food and Drug Administration (FDA). It takes many years of research to approve a treatment like this and that is no guarantee that they will get it right. Right now there is no money, let alone incentive, for the FDA to investigate any of these treatments. This despite the fact that if T cell immune system–stimulating treatments were available, they would offset autoimmune reactions that are a result of the loss of T cells experienced by older individuals (Prelog 2006). This loss of T cells pertains not only to anti-cancer T cells, but also to the T cells that control autoimmune reactions such as arthritis. If the onset of arthritis could be delayed in older individuals, the savings in health care would make up for the cost of the FDA investigation into these T cell immunity–stimulating supplements. A healthy immune system does mean longer life and lower health care cost. Getting cancer is one thing, but having to suffer from the effects of an autoimmune reaction is another. The two diseases are linked and the solution to one may bring out a cure for the other.

Why Does Cancer Involve Autoimmunity and How Does the T cell Immune System Try to Stop It?

As we have said, the T cell immune system faces a dilemma when it responds to cancer. Sometimes, the cancer peptides presented to it are truly foreign (non-self) or tumor antigen–derived, but at other times, they are half self and half non-self (hybrid) or neoantigen-derived. The neoantigen-derived peptide T cell response is a problem not only with autoimmunity, but also with accelerated tumor growth.

As far as autoimmunity is concerned, there are two options. The first option is two-fold. One, the T cells signal all responding T cells to "pause". This is a failsafe mechanism to ensure that T cells do not respond to embryonic proteins during pregnancy or to minor changes in the protein configurations of other cells. Two, the response to the neoantigen-derived hybrid peptide has to do with erroneous amino acid substitutions in these peptides. This simultaneously causes the T cells to send out a negative signal that results in widespread cancer-specific immunosuppression

(Edwards et al. 2010) and increased tumor growth (Katz et al. 1989). Both aspects of the first option favor continued cancer growth.

The second option is for the T cell immune response to proceed as if the peptide were truly non-self (tumor antigen–derived) although it is a hybrid. This too can result in an autoimmune response because of the mutation rate associated with the genes of the T cell receptor. This issue will be addressed later, but for now it is apparent that the T cell response to a cancerous cell has an inherent autoimmune quality about it.

Immediately, other T cells called T regulatory cells or Tregs (Sakaguchi et al. 2001) detect an autoimmune response. Tregs are a subset of CD4 T cells. They are created from activated T cells (antigen-specific and non-specific) when certain genes (FoxP3) are turned on that allow these cells to stop the activity of any T cell responding in an autoimmune fashion (Walker et al. 2003). Unfortunately, when Tregs detect an anti-tumor T cell response, they behave as if the response was directed at self and shut it down. In many cases this also allows the tumor to progress to a stage where it is almost impossible for the anti-tumor T cell response to destroy it. Scientists have known about these cells for some time now. However, it is not possible at this time to differentiate between a Treg that controls an anti-pancreatic (insulin-producing cell) T cell response from a Treg that controls an anti-cancer T cell response that has the autoimmune properties we have described. At this point, stopping all Treg activity is unrealistic because Tregs prevent a host of autoimmune responses that, if allowed to proceed, would certainly lead to the death of the individual. Here lies the problem, and the possible solution is dependent on this identification. If just those Tregs that inhibit anti-cancer T cells could be identified, their temporary removal would allow for many more of the cancer cells to be destroyed. This issue will come up in Chapter 11, which is devoted to Tregs.

Consequences of Preventing Anti-cancer T cells from Inducing Autoimmunity

The T cell regulatory arm of the T cell immune system is the key to controlling cancer-induced autoimmunity. However, this control also includes restraining T cells responsible for killing the cancer. Abolishing a response to normal tissues diminishes an important life-saving function of the T cell immune system. Completely removing the T cell regulatory limb of the T cell immune system would have disastrous effects, far beyond just cancer-induced autoimmunity. The effects would be widespread, involving many tissues such as connective (arthritis), pancreatic (diabetes), and neural (multiple sclerosis) tissues.

One solution to this dilemma is being examined in mice, where the Tregs are temporarily paralyzed or removed, allowing for the full response (anti-cancer and autoimmune) to proceed for a short period of time. A method called "pulsing the response" has shown some promise because of the way the T cell immune system handles cancer immunity. As the anti-cancer response ensues, a memory T cell pool is formed. Once the cancer is diminished to a certain point, the active T cell attack stops and what is left over is the memory T cells. Just prior to this point, Tregs can be released to control autoimmunity. If the cancer begins to grow, the memory T cells react quickly to kill the cancer before Tregs can stop them. In addition to this, it has been determined that the autoimmune T cell responses do not engender the same memory T cell response as the anti-cancer T cell response does (Sutmuller et al. 2001).

This is only one of the ways it may be possible to reduce the deleterious effects of cancer-induced autoimmune reactions. Identifying separate populations of Tregs that restrict only T cells that cause cancer-induced autoimmunity or anti-cancer T cells would be ideal. If it were possible to remove Tregs that restrained just anti-cancer T cells, but not autoimmune T cells, the cancer could be killed quickly. In addition, there would be no cancer-induced autoimmune reactions. Efforts are underway to do just that, since this is probably the key to not only killing the cancer, but also eliminating cancer-induced autoimmunity. This would entail a major breakthrough concerning the T cell immune system. These Tregs seem all the same but, in fact, some of them go to sites of specific autoimmune reactions as if they know where they are going. Despite this apparent dispersion of Tregs, there has so far been no identification of separate Treg populations that migrate to specific locations. If and when that breakthrough occurs, not only will anti-cancer T cell immune responses benefit, but also the side effects of cancer-induced autoimmunity will disappear.

Chapter 8

Can We Solve the Cancer Stem Cell Program and Cure Cancer Using T cell-based Immunotherapy?

When a cancer is formed, the cells are reduced to their simplest form. These cells are primordial. DNA damage sends the cells back to a more primative state. The cells are no longer lung cells, liver cells, or skin cells, but a form of stem cell sent back in time to the embryonic environment. Once back there, the cells have a new mission. Differentiation of a cell is its ability to change into another type of cell (Schjeide and De Vellis 1970). For the first time, cells have the ability to differentiate at an incredible rate. This differentiation is sometimes fueled by the immune system. The T lymphocyte is the major enforcer of the cellular immune system because of its ability to attack any foreign invader, including cancer cells. T cells recognize some of the cells and kill them, but what is left over may not be seen by the T cell immune system. The tumor mass has several different options when it comes to evading the immune system. The first two are obvious: stopping the production of recognizable cancer proteins and stopping the production of the self-proteins needed by T cells to target cancer-specific proteins expressed by the cancer. This continual differentiation allows the cancer stem cell to thrive in the face of imminent attack by the T cell immune system (Schatton and Frank 2009).

It is a dynamic process where the T cell immune system copes with these and many other changes the cancer stem cells force on the system in order to take over the person's body and cause death. In this chapter, we first describe the stem cell nature of cancer cell differentiation. We also demonstrate how the immune system can inadvertently aid cancer stem cell differentiation. We then discuss the effects cancer stem cell differentiation has on the T cell immune system. Finally, we describe possible experimental procedures that can address the cancer stem cell phenomenon and lead to effective immune-based therapies.

The Stem Cell Nature of Cancer Stem Cell Differentiation

To understand the stem cell nature of cancer, it would be prudent to explain how a normal cell becomes cancerous using a simple example. Cancer-causing agents such as ultraviolet light, chemical carcinogens and some viruses can cause breaks in the DNA of normal cells. The cell's DNA repair proteins can swiftly repair those breaks. However, when multiple breaks occur, mistakes are sometimes made. These mistakes are called mutations. When mutations occur in the genes that control cell division, a cancer cell is formed. Cancer can form in almost any cell in the body. The DNA of these rapidly dividing cells then begins to recombine in ways that are inappropriate. It tends to revert to a more immature primordial form resembling the DNA of embryonic stem cells, which have the ability to become any cell in the body given the proper signals. The only exception is that the cancer cannot be commanded to stop dividing, unlike the embryonic stem cell (Kiessling and Anderson 2003).

How the T cell Immune System Affects the Pathway of Tumor Stem Cell Differentiation

The T cell immune system sees a cell (cancer) that is not supposed to be there, then attacks and kills it. The attacking T cells do not know that there are cancer cells hiding in the tumor mass that are ready to grow out and kill the host. Once the T cells have eliminated all the cancer cells they can recognize, the ones left behind begin to grow out. These cells are invisible to the immune system. They begin to grow faster and faster because for the moment there is nothing to stop them. Some scientists believe that there is only one stem cell present in the cancer population. However, this is difficult to determine because the cells are continually dividing. As the cells grow, their stem cell nature allows them to evolve into even more dangerous adversaries (Frank et al. 2010). Our immune system can handle almost every type of foreign threat, but cancers like this present multiple problems that go beyond just not being recognized.

Problems for T cells Faced with Tumor Stem Cell Differentiation

What makes cancer difficult to fight is that just before a cell becomes cancerous it is a normal cell. The immune system is programmed not to respond to normal cells (self). The potential cancer cell may have been ignored by the immune system for decades. T cells then have a major problem. They eventually begin to eliminate the cancer cells that they recognize (non-self). The stem cell nature of the cancer cell causes it to produce new cancer cells expressing mutant proteins that are very similar to self-proteins, but not quite self. The T cells get confused. They begin to respond, but they are actually responding to self.

The immune system applies pressure on the cancer stem cell population. Despite these efforts, the cancer cells change and evolve. The immune system can actually select highly lethal cancer variants that are invisible to the immune system. This process is called "immune selection". Immune selection can cause even more subtle problems for the T cell immune system.

Some cancer stem cells can send signals to other cells (macrophages) of the immune system and cause them to secrete a chemical called indoleamine 2, 3 dioxygenase (IDO) that paralyzes all T cells in the area of the cancer cell mass. This happens because T cells need an essential amino acid called tryptophan in order to proliferate. IDO breaks down the amino acid and causes a type of generalized T cell tolerance. As the cancer stem cells evolve and change, they eventually reach a state where they can secrete anti-inflammatory proteins. This naturally causes macrophages to secrete IDO. The T cells take some time to recover, so some of them wind up in other areas of the body still unable to respond to anything. This type of broad-based T cell suppression is harmful to the body, since even a simple bacterial infection can be deadly (Chen et al. 2008). These are just a few of the ways in which an ever-changing cancer stem cell subverts the T cell immune system. The changes it displays are a direct result of how the T cell immune system works. The cancer stem cells take advantage of the self/non-self rule, induce autoimmunity, and initiate anti-inflammatory responses (Galgani et al. 2009). Methods must be developed to understand the tumor stem cell program in order to predict when these changes will occur and how to subvert the ability of these cells to change at such a rapid rate.

An Experimental Approach to Cancer Stem Cells toward Effective Immune-based Anti-cancer Therapies

The concept of cancer stem cells is new. In contrast, for years it had been recognized that the T cell population was multi-clonal. That means it is made up of hundreds of separate clones. Each clone has at least one, but sometimes two, specific immunological targets. Their numbers are low and sometimes less than 1 percent of the total population. However, when they see the specific target their number expands and can reach 50 percent. After the immunological target is eliminated by the T cell clone response, the levels of that particular clone return to normal, leaving some memory T cells behind. The T cell immune system works this way because it's the best way to prepare for an invading bacterium, virus, or cancer

(Rothenberg et al. 1992). Strangely, the cancer stem cell works in the same way, but in reverse.

The clones of the cancer stem cell population begin to expand one at a time or in multiple numbers. Some are completely decimated and others are reduced in numbers by the T cell response. For each clone of cancer cells that is destroyed, there is another one waiting to grow. During this process, the cancer cells change and evolve (Frank et al. 2010). They confuse the T cells because of a biological program embedded in their DNA. The cancer stem cells could, without immunological pressure, overcome a person's body, resulting in death. However, immunological pressure occurs slowly. Sometimes, the slow development of cancer cell differentiation due to this pressure results in a highly lethal clone of cancer cells. This clone of cancer cells is what inevitably leads to the death of a person.

Now that we see the scope of the problem, a solution must be arrived at that takes into account the variables set forth by the differentiation process. Since selective pressures of the T cell immune response drive this process, a model system can be developed to mimic the process of forced cancer stem cell differentiation. The model must identify two things: first, the measurable death of cancer cells at each step, and second, measurements of the resulting change in the character of the remaining cancer stem cells.

A model must be developed that can isolate the variables of the system in such a way as to make it easy to manipulate them and identify which one affected the system. An *in vitro* (in the laboratory) system is preferred over an *in vivo* (in the body) model because of its simplicity. Sometimes, biological reactions that occur in the body are influenced by unknown factors, making it difficult to pinpoint the cause of a given response. The information derived from the model must be precise because it must be replicated to verify its significance. In other words, the experiments must be simple and easily reproducible. This ensures that the information is valid and can be used as a basis for subsequent experiments.

With that in mind, an *in vitro* system involving the T cell immune system would be a population of multi-clonal anti-cancer T cells produced in the laboratory directed against cancer cells derived from the same individual. The ability of these T cells to kill the cancer cells can be easily measured in the lab. Populations of T cells would be chosen that demonstrate the maximum amount of killing.

As far as the cancer stem cell population is concerned, it is possible to establish cancer cell lines in the lab from patient biopsies that approximate the original make-up of the cancer stem cell population. Once these cell lines are established and expanded, multiple samples are frozen in the early stages of cancer cell culture. This ensures that each population of cells thawed and grown has exactly the same proportions of cancer stem cell clones. With the advent of DNA technology it is possible to determine which genes the cancer cell line expresses. This is called DNA micro array analysis (Schena 1999). Cancer cells from the initial culture are analyzed for expression of known genes in order to develop a base line. This base line is important. What happens next will expose the cancer stem cell program.

Killer T cells are incubated with the cancer cells. Many of the cancer cells get killed. The survivors must be analyzed for expression of the same genes. After a short culture period, the gene expression changes because some of the cells expressing them are missing. At this point new genes become expressed because other stem cell clones are growing. This data (gene expression) should be categorized and examined for possible effects of the expressed protein products on not only the immune system but also the body in general. This process of killing and gene expression should be repeated until noticeable patterns appear. Also, similar data from other patient T cell/cancer cell line experiments should be compared so that a universal pattern can be detected. From data like this it should be possible to design treatments to circumvent the changes before they happen.

This is only one approach to unlocking the cellular code or program that cancer stem cells use to overcome their host. These cells are very adaptive in that they seem to be able to circumvent the best efforts of our immune system. Much more research must be done to fully understand how this happens. Techniques must be developed that can decipher the actions of cancer stem cells so that suitable treatments can be developed that stop the cellular progression that for some people lead to death. Our technology and knowledge of the immune system has progressed to the point where one day each change a cancer stem cell goes though can be predicted and countered by immunotherapeutic measures. Hopefully, this will result in an immediate increase in the cure rate for some of the most difficult types of cancer.

Chapter 9

Translational Immunology and Its Impact on Anti-cancer T cell Immunity and the Development of Immunotherapeutic Treatments for Cancer

Translational immunology is an emerging field and represents the combination of several scientific specialties. Translational biology, as applied to immunology, is the study of intercellular biochemical pathways within cells that allow cells to function in a specific manner. Many of these pathways intersect and are influenced by external stimuli, as when a cell's receptor, say, the T cell receptor or TCR, comes in contact with its ligand (peptide antigen/MHC complex). This interaction induces a series of biochemical changes within the T cell to signal it, for example, to divide, secrete lymphokines, or differentiate into another type of T cell (Birmingham 2002). Because this is an emerging field of study, many basic T cell immunologists, like me, have been waiting in the wings, until the science of intercellular pathways started to make sense. Most people have had difficulty understanding this subject, but it is something we need to know.

Translational biology has been studied for decades, but in the past few years some of the discoveries concerning immune cells have been astounding. As you read through some of the later chapters you will gain an appreciation of the biochemical processes that enable the cells of our immune system to function. At times, the biochemistry of the processes can seem complicated, but there is a sort of symmetry to it. The T cell immune system, including all of the supporting immune cells, depends heavily on the biochemical changes that occur within each cell, especially during a T cell response to a tumor.

In the end, these biochemical pathways direct immune cells to do what they do in a specialized manner. The field of immunology is ripe with opportunities to alter immune cell behavior to the point where, if a cell is behaving badly, it may be possible to interrupt the cellular pathways that induce the unwanted behavior and correct them. Farfetched as it may seem, in the future there will be an entirely new generation of chemical compounds that alter cellular functions to produce any number of cellular behaviors. This is the new frontier in immunology. One just has to examine the immunological research that is being done to realize that understanding translational immunology means controlling the immune system. Using this tool, it may be possible to not only change the way the T cell immune system sees cancer, but possibly change the way cancerous cells present themselves to that system.

Translational immunology is complex because it involves biochemical changes that occur inside a cell and that happen almost immediately after the cell has encountered a receptor or ligand. These events are important, given their role in determining how immune cells function. In this chapter, we first examine the biochemical intercellular changes a T cell goes through when it encounters an antigen, which its T cell receptor can attach to. Next, we briefly examine the intercellular changes that occur in supporting immune cells such as DCs and NK cells that allow them to participate in the anti-tumor T cell response. Finally, we examine immune cell intercellular pathways and how they may have a positive impact on the development of immunotherapeutic treatments for cancer.

Intercellular Signaling as a Result of T cell Receptor Engagement

The T cell receptor is a series of proteins (molecules) spanning the cell membrane of the T cell. T cells are quite complex in their behavior. This complexity is reflected in the many surface receptors (locks) and ligands (keys) they express on the outside and inside of the cell membrane. Many of the proteins that appear on the cell surface have what is called an intercellular domain. The domains of these surface molecules extend a short distance within the cytoplasm of the cell. So the reality is that these proteins have two faces. One is outside the cell membrane and the other one is inside. However, these two parts of the same protein have different functions. The outside portion can function as a receptor or as a ligand. Its job is to interact with other cells and, in the case of the T cell, with foreign antigens such as cancer-related peptides. The intercellular portions of the membrane proteins have a different function. Their job is to communicate with other proteins within the cell, which directs them to eventually address the DNA of the cell so the DNA can produce specific proteins as a part of its response to this outside stimuli.

The TCR is a complex of proteins that come together when the cell interacts with peptide antigens. This complex is composed of TCR-alpha, TCR-beta, and three varieties of the CD3 proteins (epsilon, delta, and gamma). These CD3 proteins not only span the cell membrane, but protrude from the cell surface. They are needed to assemble the TCR-alpha and -beta chains on the cell surface. Inside the cell, two CD3-zeta proteins are associated with the ends of both the TCR-alpha and -beta proteins on the inside of the cell membrane and protrude into the cytoplasm. Depending on the type of T cell, a CD4 or CD8 protein is necessary for the contact between the antigen-presenting cell (APC) and the T cell. If this is a CD4 T cell, there is an affinity for MHC class II (HLA-DR, DP, DQ, or DM) proteins because the CD4 protein (ligand) attaches itself to the MHC class II protein (receptor). A CD8 T cell has an affinity for MHC class I proteins (HLA-A, B, C) for the same reasons (Lefranc and Lefranc 2001). Now, the TCR is prepared to attach to the presented peptide. Chapter 2 describes how the peptide antigen is arranged on the APC surface. So now you have the protein players and their roles in the detection of non-self peptides. One more thing to remember is that the TCR-alpha and -beta proteins are configured so as to conform to the peptide-MHC protein complex on the surface of the APC.

All this is slightly complicated, but what happens inside the cell when the foreign peptide-MHC complex on the APC engages this TCR-complex is more so.

Contact is now made between the TCR-complex and the MHC-peptide complex. The four CD3 subunits have a sequence motif (similar amino acid repeats) called immunoreceptor tyrosine-based activation motifs or ITAMs and these serve as protein tyrosine kinase substrates. ITAMs are phosphorylated (PO_4 is added to them) by tyrosine kinases from the Syk family (Zap-70) and the Src family (Lck). Phosphorylated ITAMs can serve as binding sites for SH2 domains of other protein kinases that basically move off and begin to phosphorylate other proteins in a cascading manner (Sinclair 2000). This description focuses on the TCR-CD3 complex, but there are several other steps that involve the engagement of either the CD4 protein or the CD8 protein. These engagements result in a similar cascading effect, which involves other phosphorylation events with other kinases. Those events should be examined as well, but they will not be detailed here. This is to minimize the confusion due to the array of events all occurring at the same time. However, as these events occur, the reactions move close and closer to the DNA of the T cell. There are at least four intercellular molecules that are important in this regard. The first is p38 MAP kinase, then NFAT, which p38 can activate. NF-kB is another, which is right outside the nucleus, as is NFAT. Then there is AP1, which can be activated within the nucleus by Erk and JNK2. Three of these can directly signal the activation or inhibition of gene expression within the nucleus itself. As mentioned before, the results of this gene activation or deactivation comprise the response T cells go through when they are activated (Marks et al. 2009). The changes in gene expression in activated T cells determine the behavior of the stimulated cell. They can be instructed to kill a target, divide, secrete specific lymphokines, and serve other functions. All these activities determine the type and strength of the T cell response to a tumor and all are induced by intercellular signals relayed down pre-determined pathways within the cytoplasm of the responding T cell.

Intercellular Pathways Used by Other Immune Cells that Support the Anti-cancer T cell Response

Dendritic cells

After the T cell series, my next favorite immune-cell class comprises the cells of the myeloid lineage. Of particular focus will be the dendritic cell (DC). Since this cell is, in my opinion, the second most powerful cell involved in the anti-cancer T cell response, it is very important to understand the significance of the intercellular pathways that occur within DCs. Just like T cells, DCs go through morphological (cell type) and phenotypic (functional) changes in response to outside stimuli. The changes in these intercellular pathways control important aspects of the DC's behavior. Depending on the stimuli or the strength of such a signal, the intercellular events that take place within the cell can change it permanently. For instance, if a DC stops in a particular location, say the skin, it changes into a Langerhans cell (LC) for the rest of its life. Coincidentally, LC is the longest living DC-like cell in the body. It can live for over two years in the skin, whereas organ-specific DCs like the Kupffer cell in the liver need to be replenished at a much higher rate in a shorter period of time. In a human skin or limb graft, the donor's LC can be detected for over two years. The morphological changes (from DC to LC) are preceded by intercellular events induced by the environment the DCs find themselves in, which in this case is the epidermis (skin). This change occurs in response to the secretion of colony-stimulating factor-1 (CSF-1) and transforming growth factor-1 (TGF-B) by keratinocytes in the skin (Merad et al. 2008). Once these cytokines act on the newly arrived DCs, intercellular pathways are activated. These are related to the inhibition of a transcriptional repressor called GFi-1 (from growth factor independence gene-1), which regulates STAT3 that can activate DNA promoters (Melillo et al. 2010). This selective promotion of genes results in morphological and phenotypic changes characteristic of the LCs.

Receptors that cells possess initiate the activation of intercellular pathways that can either activate or deactivate cellular functions. Several of these events occur in DCs. The first involves a ligand on an activated T cell called lymphocyte activation gene-3 protein or LAG-3 and the receptor on the DC is the MHC class II protein. Because LAG-3 is part of the CD4 family of ligands, it has an affinity for MHC class II proteins. When the LAG-3 comes into contact with the MHC class II protein on the DC cell surface, intercellular changes begin to occur on the inside of the cell membrane (Liang et al. 2008). Several complexes are attracted to the intercellular portion of the MHC class II protein. These are lipid rafts (Cho 2010) and Fc receptor domains, which contain ITAMs, the protein kinase substrates. Unlike the positive situation with the T cell receptor, phosphorylation of ITAMs in this situation induces a negative signal because ERK and SHP-1 are recruited. The recruitment and phosphorylation of ERK is particularly inhibitory because it has been shown to inhibit the ability of DC to activate Th-1 CD4 T cells by secretion of IL-12. This is called an ITAM inhibitory pathway and the end result for the DC is that it inhibits DC activation and suppresses maturation of these cells (Liang et al. 2008). Both these inhibitory actions prevent DCs from inducing a productive anti-tumor T cell response. It was found that this activity, due to the cross-linking of MHC class II by LAG-3, could be inhibited by methyl-B-cyclodextrin, a lipid metabolism inhibitor (Cifuentes et al. 2008), which interferes with a protein's association with lipid rafts that SHP-1 is attached to. Furthermore, ERK had to be phosphorylated as an upstream event in order for SHP-1 to associate with the lipid rafts because once ERK was inhibited with a compound called PD98059, the association did not occur (Liang et al. 2008).

Another important DC receptor is the B7 (CD80) protein. The natural ligand for this receptor is called PD-1, which is expressed on activated T cells (T regulatory cells). One study revealed that cross-linking B7 proteins resulted in not only activation of DC functions such as antigen presentation, but also their ability to "reprogram" T regulatory cells (Tregs). The Tregs were converted into an autoimmune effector cell (Th-17 CD4 T cell) capable of inducing autoimmune reactions and breaking established self-tolerance as well as killing target cells, much like CD8 effectors can. This activity, induced by cross-linking B7, has several wide-ranging implications related to the anti-tumor T cell response. First, many Tregs that inhibit tumor-specific T cell responses were themselves at one time anti-tumor T cells with an anti-tumor TCR. The reason for this is that some anti-tumor T cells are converted to Tregs by DCs during the initial anti-tumor T cell response. Second, once the Tregs are converted to autoimmune effectors, they not only induce inflammation at the tumor site, but also attack and kill the tumor and can disarm Tregs trying to protect the tumor from attack. However, there is another aspect of Treg to autoimmune effector activity that may be harmful to a cancer patient (Radhakrishnan et al. 2008a).

The Tregs are tumor-antigen–specific (Wang et al. 2005) before their conversion to effector cells. There are Tregs present in a tumor mass that are non-tumor–specific, such as naturally occurring or "innate" Tregs. When these Tregs are converted to autoimmune effectors they can break immune tolerances not associated with the tumor. This can result in neural damage, death of islet cells (insulin-producing cells), colitis, or even lupus. All these autoimmune diseases have been found in cancer patients with an active anti-tumor T cell response. This is referred to as autoimmune toxicity (von Euw et al. 2009). It is a serious problem for cancer patients who are being treated with antibodies that attack Tregs in a system-wide (infusion) application during clinical trials. When the anti-Treg (anti-CTLA-4) antibodies are used, they increase the presence of the same Th-17 autoimmune effector cells. The autoimmune effectors are generated from innate Tregs that have little expression of the CTLA-4 activated T cell antigen. Therefore, in response to the loss of tumor-specific (activated) Tregs, the DCs begin to convert innate Tregs to autoimmune effector cells, which cause autoimmune toxicity (von Euw et al. 2009).

The intercellular events that induce this activity involve the phosphorylation of a transmembrane-anchored protein adapter called DAP12, which contains an ITAM within its cytoplasmic domain. Because the adapter has an aspartic acid in the transmembrane portion, it associates with positively charged amino acids of other transmembrane proteins. In this case, it is attracted to the B7 cross-linked protein transmembrane domains. This action with DAP12 activates the protein kinases Syk and phospholipase C (PLC-g). This Syk to PLC-g activation leads to increased DC activation in the form of antigen presentation, but it also induces DCs to secrete IL-17, INF-g, and TNF-A, which induce the conversion of Tregs to these autoimmune effector cells. This conversion of Tregs can be harmful if the converted Tregs are non-tumor-specific. In this case it would be prudent to inhibit the conversion, especially under conditions that generate Th-17 effector cells, like using anti-Treg (anti-CTLL) antibodies. There is an inhibitor of Syk called piceatannol, which blocks this pathway and in doing so prevents DCs from converting Tregs into Th-17 autoimmune effectors. However, piceatannol does somewhat inhibit antigen presentation by DCs, but at least the DCs will not be able to induce life-threatening autoimmune reactions through the generation of autoimmune effector cells (Radhakrishnan et al. 2008b).

Another ligand-receptor DC activation event involves an intercellular pathway that can either induce DCs to make Tregs or make them secrete Th-1 lymphokines, which induce a productive anti-tumor T cell response. This has to do with the toll-like receptors (TLR) all DCs possess. First, a little about what they do when engaged by their ligands and the role TLRs play in DC immunology. Next, a look at how intercellular pathways control the end result of this ligand-receptor interaction. The TLR family of receptors is called pathogen-related molecular pattern-recognizing receptors, of which there are eight. Five of these receptors (TLR 1, 2, 4, 5, and 10) recognize bacterial products, while the remaining three (TLR 3, 7, and 9) see viral nucleic acids. Some of the natural ligands of these receptors are lipopolysaccharides, a bacterial product, and unmethylated CpG DNA motifs (free-floating DNA from disintegrated bacteria). Once the TLRs are engaged by these ligands on the DC surface, a reaction starts that results in the DC secreting IL-10 and TGF-B (Groschel et al. 2008). This lymphokine and cytokine induce the conversion of T cells into Tregs (Lombardi et al. 2009). It happens because the DCs anticipate a massive infection, where T cells sometimes turn into autoimmune effectors either by conversion or mutation of the TCR. This is a way for the immune system to control that aspect of rapid T cell proliferation caused by a massive bacterial or viral infection.

As far as intercellular events are concerned, p38 MAP kinase signaling is involved. The signaling involves phosphorylation of p38 and the subsequent secretion of Treg-inducing proteins (IL-10 and TGF-B). Once several inhibitors of p38 phosphorylation are used, the DC activated by TLR ligands cease to produce both IL-10 and TGF-B. Because TLR engagement induces DC activation and the p38 pathway is blocked, the DCs begin to secrete IL-12 and INF-g. Both of these induce Th-1 T cells. As mentioned before, p38 is a downstream activator and it can activate NFAT. NFAT sends a signal directly across the nuclear membrane to induce a series of events that will result in anergy, which results in Treg formation by DCs. However, when this signaling by p38 is blocked, NFAT cannot send the inhibitory signal. Blocking p38 just reroutes the signal, which can involve several alternative pathways. One of these is the JNK2 to AP1 (inside the nucleus) pathway. This pathway leads to proliferation and differentiation steps that DCs must undergo in order to secrete Th-1 inducing lymphokines (Yi et al. 2002).

NK cells

Let's turn our attention to a cell that is not a part of the T cell lineage but plays a major role in the anti-tumor T cell response. This is the NK cell. Like most immune cells, it is born in the bone marrow. The NK cell has a special function with respect to T cell immune response to cancer. It can detect any cell that has irregular proteins expressed on its surface (mostly glycoproteins). The NK cell is part of the "immune surveillance system" and, since it is not a T cell, it does not go through the thymus. Unlike T cells, NK cells develop no memory of previous encounters. Their sole job is to find a suitable target and kill it. They do this much like CD8 killer cells do, using perforin. Perforin is an enzyme stored in granules of the cytoplasm of effector T cells, DC, and NK cells; when released, it bores a hole into that cell. The content of the target cell's cytoplasm pours out of the cell, causing it to die. The reason NK cells are important to the anti-tumor T cell response has to do with antigen presentation. After an NK cell attack, many dead tumor cells are left in its wake. Macrophages come in and devour the cellular debris and process the tumor antigens into immunogenic peptides. Anti-tumor CD4 T cells begin to respond to these tumor peptides and a classic anti-tumor T cell response ensues. Natural killer cells have a special relationship to DCs in that, to some extent, DCs can activate and release them from their home in regional lymph nodes. That activation and release is governed by intercellular events that occur as a result of DCs making contact with the NK cell (Moretta 2003).

When DCs make contact with NK cells, they produce IL-2. IL-2 activates the NK cells to kill target cells. This is an autocrine reaction, which means that the lymphokine released by the cell is affecting primarily that cell. IL-2 activation mobilizes phosphatidylcholine-specific phospholipase C or PC-PLC from the cytoplasm into the periphery of the cell. While this is happening, intercellular calcium is increased and protein kinases are activated. This process concentrates granules containing the perforin used to kill target tumor cells. The granules move along microtubules after IL-2 activation, and PC-PLC enzyme activity makes this possible. Inhibitors of the PC-PLC enzyme reduce the ability of NK cells to not only become activated, but also kill tumor targets. This intercellular enzyme is important to the ability of NK cells to activate and kill tumor cells. Without the ability of NK cells to kill tumor cells, the T cell immune system may not be alerted in time to mount a response. The regulation of these perforin granules in NK cells by this enzyme is an essential function, and without the translocation of PC-PLC, it cannot happen (Ramoni et al. 2001).

NK cell receptors initiate signals through ITAM-containing adapters such as Fc domain and DAP12. The adapters send signals directly to the NF-kB complex. This complex after proteolysis sends a positive signal to the DNA. However, these adapters, in order to activate NF-kB, must first recruit caspase recruitment domain 9 or CARD9. This is a unique pathway for NK cells because it controls NK cell killing as well as the ability of NK cells to secrete lymphokines and cytokines. It apparently is used by five separate receptor-ligand systems in NK cells and only two in T and B cells; this separates the NK cell lineage from the other immune cells (monocyte-macrophages, Polymorphonuclear cells (PML), and DC) that play a role in generating an anti-tumor T cell response. The finding is significant because protein kinase C is involved and inhibitors of these kinases alter only the NK cell's ability to release lymphokines and cytokines (Hara et al. 2008). This is important because NK cells play a role in inhibiting graft vs. host disease, a potentially dangerous side effect of bone marrow transplantation for leukemia. The treatment of immunotherapeutic NK cells with inhibitors of Protein Kinase C (PKC) would prevent this unwanted activity. This would allow the NK cells to control graft vs host disease while the transplanted cells attack the leukemic cells.

The NK cell is different from all other immune cells because its cytoplasm is packed with pre-arranged proteins. The difference has to do with the density gradient, usually made up of different concentrations of sucrose (sugar); the most dense sugar is at the bottom of a centrifuge tube and the least dense is at the top. If you take whole blood and put it on the top of this gradient and subject the tube to centrifugal force, the red cells (blood cells) will be at the bottom. They have no nucleus or cytoplasm so they are all protein and dense, which means they fall. This gradient separates white blood cells on the basis of the ratio between the nuclear material and the cytoplasm. T and B cells segregate together and monocytes and macrophages do as well, but NK cells are separate. Natural killer cells are the "bullies" of the immune system. Very few immune cells aside from DCs have anything to do with them. T and B cells just get out of their way and wait for the NK cell to do its job. Natural killer cells use an alternative method to attack tumors and virally infected cells. Their recognition of these unwanted cells is a bit unorthodox; nonetheless, they are an essential component of the anti-tumor T cell response. In a sense the intercellular pathways that govern this cell's behavior are far more complicated than in any other immune cell. The ability to grow these cells in the lab has just recently been discovered. The NK cell is probably the last cell in the immune system that scientists have been able to figure out.

Recently Discovered Immune Cell Intercellular Pathways and Their Potential in Development of Immunotherapeutic Treatments

Let's look at what is new in translational immunology. Pathways involving T cell activation are the rage. They may alter the way we perceive the anti-tumor T cell response. If we could change the way T cells respond to tumor cells above and beyond their natural role, the impact on cancer treatment would be phenomenal. T cells and NK cells carry out "immune surveillance" by making contact with various cells in the body. This process never stops during our lifetime. Those cells make contact through the "immunological synapse" (Valitutti and Dupre 2010). The immunological synapse is the interface between two cells, one cell being the T or NK cell and the other, say, a tumor cell or a cell that may be developing into a tumor. One of the most recent findings in translational immunology is that information is passed from the tumor cell to the lymphocyte. This information is in the form of an oncogene called H-Ras. Now, H-Ras is called a G-protein and is an integral part of the ability of lymphocytes to function. It is a part of the multiple intercellular pathways mentioned (pre-MAPK signaling) and is an upstream activator (phosphorylation) of Erk, which activates AP1 toward the mobilization of genes in the DNA that induce positive activation and differentiation of the cell. As a matter of fact, Erk is intimately involved with the induction of NK cell cytotoxicity, through the phosphorylation of AP1. Therefore, transfer of H-Ras to an NK cell by a tumor cell almost guarantees that the NK cell will kill the cell that transferred the H-Ras.

Lymphocytes (T cells and NK cells) use this information to decide what action to take during their encounter with a tumor cell or an altered normal cell. The imported H-Ras is directly associated with the lymphocyte's intercellular pathways and if enough is accumulated, the cell will respond by killing the suspect tumor cell. This signal, received by the lymphocytes, can alert them as well as the rest of the T cell immune system that a foreign entity has just been created. However, tumors use similar mechanisms to induce non-responsiveness in lymphocytes, but this method of extracting an oncogene from a cancerous cell usually results in the death of that cell by the T or NK cell. This process is an essential part of the natural immune system. Just as the "innate" T cell populations protect us with programmed genetic information in their TCR, this system is important in initiating a T cell response to a tumor. This transfer of the oncogene protein H-Ras from suspect cells to lymphocytes can possibly be exploited and used to design new anti-cancer therapies that would make it easier for the T cell immune system to attack a tumor. The idea is that if a tumor is detected early, there is a greater likelihood that the T cell immune system can destroy it (Rechavi et al. 2007).

A recent investigation revealed a complex series of intercellular events that occurs in all T cells. It involves the T cell's ability to move from one place to another through tissues and organs of the immune system, such as lymphatic vessels and lymph nodes. Much of this movement involves leukocyte function-associated antigen-1, or LFA-1. It is mentioned in Chapters 10 and 11, on dendritic cells and regulatory T cells respectively. This ligand, found on T cells, has its receptor (ICAM-1) expressed by cells of the myeloid series, such as the DC. The motility aspect of the issue is related to how rapidly Tregs move toward activated DCs compared to the speed of naïve antigen-specific T cells that need to reach the DCs to start a productive anti-tumor T cell response. The interaction between the ligand and receptor blocks naïve T cell access to DC and also prevents the DCs from maturing fully. It does this by down-modulating the B7.1 (CD80) and the B7.2 (CD86) proteins expressed by DCs and needed by the same naïve T cells to initiate T cell proliferation. Because this activity cements these two cells, it is important to deal with this phenomenon in the context of immunotherapy. LFA-1 expression plays a major role in T cell motility. Controlling this activity and understanding the pathways generated inside the T cell via LFA-1 can be an important tool to control the fate of naïve tumor antigen–specific T cells (Onishi et al. 2008).

LFA-1's contact with ICAM-1 affects the DC as well as the T cell expressing LFA-1. This is called a bi-directional pathway, affecting both cells, but in different ways and using different intercellular mediators. This activity within the T cell is called an "inside-out" signal. LFA-1 is highly expressed on activated T cells (Tregs), but it is in a low-affinity state in resting T cells (naïve). One of the many events that change LFA-1 into a high-affinity state is the engagement of CD4 with MHC II or CD8 with MHC I, both receptors of these T cell ligands. Control of T cell motility is an essential function to include in future anti-cancer immunotherapeutic design. Once this process begins, LFA-1 is converted to the high-affinity state by CD4/CD8 engagement. Lck is activated and phosphorylates Zap-70, which in turn does the same thing to an adapter called SLP-76. Two more steps lead to an increased cytoskeleton activity. The cytoskeleton of the T cell controls its ability to move. The motility of T cells rapidly increases as a result of this intercellular pathway activation (Horn et al. 2009).

Many other pathways intersect with this one, but for simplicity's sake this is the basic pathway. This finding may allow for the augmentation of naïve tumor-antigen T cell motility. These cells need to find activated DCs before Tregs do. Increasing naïve T cell motility may increase the likelihood that activated tumor antigen–pulsed DCs, prepared for immunotherapy and infused back into a cancer patient, will immediately make contact with these faster naïve T cells. Potentially, both DCs and fast naïve T cells can be prepared. The technologies to do this probably already exist, such as anti-tumor TCR transfection of naïve T cells, and the introduction of Lck into naïve T cells by the same method, which will increase T cell motility. The idea of treating cancer with immunotherapeutic protocols like the one proposed here brings hope that immunotherapy can cure some of the more difficult types of cancers.

This chapter, which may seem difficult to understand, describes only part of what is happening in the world of translational immunology. Many of the functional activities directed by these cellular pathways are at the root of basic immune cellular behavior. If we can understand these pathways, we may be able to develop cures for cancer. Most of what is described here is directly related to how immune cells (T cells, NK cells, and cells of the myeloid lineage) are directed by external stimuli related to the presence of a tumor. However, there is another aspect to these pathways that is related to those intercellular pathways used by the tumor to deal with the anti-tumor T cell response. This was omitted in order to streamline and simplify this subject, but there are aspects of tumor-associated pathway activation that are just as important as those mentioned in this chapter. What is very exciting about all of this is that these pathways can be exploited to benefit cancer patients by increasing the effectiveness of immunotherapeutic treatments. The application of beneficial changes in immune cell behavior by translational immunology, coupled with already proven therapies, will go a long way toward curing previously difficult forms of cancer. This area of research may very well hold the key to the ultimate treatment for cancer in the future.

Chapter 10

Dendritic Cells, Immunotherapy and the Anti-cancer T cell Response

In Chapter 1, dendritic cells (DCs) were featured prominently because of the role they play in the anti-cancer T cell response. Dendritic cells figure prominently in any T cell activity. Without DC involvement, the anti-cancer T cell response would be like a ship with a broken rudder. Its captain would have to struggle to keep it on course and avoid dangerous obstacles. Dendritic cells can streamline the T cell response so as to direct it to its desired target (cancer) without damage to normal tissues.

Why are DCs so important to the anti-cancer T cell response? Suppose you are observing a battle from one combatant's side, say, the cellular immune system. The other side is a developing cancer. The two armies have many soldiers with specific functions. There are the enlisted men (T cells), who are on the front lines doing the actual fighting, and then there are a series of commanders. Some of these commanders are back at headquarters and some of them are in the field with the grunts, as they are called. There are two of these field commanders. One is in back of the front line and the other is in the field where the fighting is going on.

The DC plays both these roles. Once they are created by the T cells, both types of DCs are involved. The DC that is behind the battle line makes sure the army (effector/helper T cells) is equipped with the appropriate weapons: in this case, DC antigen presentation and activation of T cells. The other DC in the battle is similar to the commanding lieutenant of a fighting unit, making sure that all the men are directing their weapons at the enemy. He also has to coordinate the fight to minimize deaths from friendly fire. Dendritic cells continually activate T cells to respond and kill the tumor, but at the same time induce T regulatory cells, which prevent anti-cancer T cells from attacking normal tissue. What cell or cells play the part of the commander at HQ? We will discuss that later, but now we will summarize our examination of DCs and the anti-cancer T cell response.

We first learn exactly what DCs are and where they come from, and also what types of DCs inhabit our bodies. We also investigate what functions these cells have aside from their role in the anti-cancer T cell response. These functions, which seem to have nothing to do with the anti-cancer T cell response, actually are quite important. Next, we examine the interactions between DCs and T cells: why and how T cells create DCs, and why the types of DCs created by T cells can hamper the anti-cancer T cell response. When we look at how DCs control T cell activity and conversely how T cells control DC activity, we can better understand the real problems associated with DC-based immunotherapy. Finally, we examine in detail how DCs are being used to treat cancer. We will look at the failure of DC-based therapy to induce cancer regression, which is largely due to the activity of T regulatory cells (Tregs). In closing, we will outline the intercellular events that occur within DC and alter their function. Also, we will examine how these pathways and the Treg-DC cellular contacts can be manipulated in a way to improve the ability of DCs to induce a productive anti-cancer T cell response during DC-based immunotherapy.

Origin, Function and Types of DCs in our Bodies

Dendritic cells come from the myeloid lineage of the bone marrow. The myeloid lineage gives rise to the monocyte and the macrophage. This lineage of immune cells is probably one of the most diverse groups of cells in our bodies. Briefly, the initial cell to emerge from the bone marrow is the monocyte. This is basically a circulating undifferentiated cell that has an immediate function in the blood, which is to congregate at sites of tissue damage or inflammation. But if called upon, the monocyte can enter the tissues and turn into a multitude of different cells that have multiple duties to perform. Monocytes that enter specific tissues differentiate and take on the appearance of the resident cells. For instance, when they take up residence in the brain they are called microglial cells. Under the microscope, one of these cells would look like a neuron to an untrained eye. In the liver, they become Kupffer cells and in the skin they become Langerhans cells. In every organ system, they all turn into one type of cell regardless of what they look like. The simple change a monocyte goes through once it stops at any given site is to turn into a macrophage. The macrophage is a somewhat more specialized cell in that it has the ability to eat debris. That debris includes bacteria, dead cells, dead viral-infected cells, and other pathogenic and non-pathogenic material. In other words, everything it comes across. Once this material is ingested it is reduced by acids in the cell into protein subunits called peptides. These peptides are associated with self-proteins and expressed on the cell surface for T cells with the appropriate receptor to respond to. This activity is called antigen presentation (see Chapter 2 for a detailed description).

Dendritic cells are basically created from either the monocyte or the macrophage when they interact with two small protein molecules called lymphokines, both secreted by the T cells. The lymphokines are IL-4 and granulocyte-monocyte colony stimulating factor, or GMCF. There are two phenotypic classes of DCs: the tissue DC, which is the subject of this discussion, and the plasmacytoid DC (pDC). The pDC mainly circulates in the bloodstream but can inhabit the liver. The standard DC is found in the blood and in almost every tissue-organ system in the body. Both these cells have antigen-presentation capabilities, but the pDC has a heightened ability to create Tregs in situations when autoimmune/inflammatory conditions are present (Lombardi et al. 2009a). This will be examined in detail when we discuss how DCs induce Tregs.

Dendritic cells tend to create homeostatic conditions within the immune system (Merad and Manz 2009). Aside from their function in anti-cancer T cell responses, they are involved in a myriad immune responses. Many of these functions (including antigen presentation) have to do with refining the T cell response to pathogenic organisms, viruses, and inflammatory responses. Because DCs can induce Tregs, they can control autoimmune reactions that, for instance, could lead to diabetes, arthritis, and lupus. For instance, if a person takes steroids, the steroids tend to alter cells slightly, such as islet cells in the pancreas. The cell membrane of these cells becomes irregular in response to the steroid. The immune system notices this and initiates an inflammatory response. This response requires that macrophages, mast cells, and a host of other cells appear at the site to protect the body from a widespread autoimmune anomaly that could threaten life. These cells don't know why this is happening and that it is just a local phenomenon, but the call still comes out. Once the islet cells change to the point where the immune system thinks they are not self, then all hell breaks loose.

The proteins made by these steroid-altered islet cells are so different from normal islet cell proteins that they register as foreign to the T cell immune system. A change in protein recognition of this magnitude results in the creation of the previously mentioned "neoantigen" by the cell in question. The immunogenic peptides now produced by the islet cells resemble normal peptides, but they are so distorted by the steroid effect that they seem foreign. As a result of the appearance of these altered proteins, a T cell response is initiated. Part of that response requires that the T cells release lymphokines, which induce the formation of DCs from the arriving monocytes and macrophages. If the signals are slight, immature DCs are formed, and they literally reach out to the neoantigen-specific T cells and convert them into Tregs (Mahnke et al. 2007). The Tregs have the ability to stop T cells from attacking normal tissue, which in this case, despite the fact that they are producing altered self proteins, is the islet cell. They come in and halt all anti-islet cell activity mediated by T cells. This prevents the islet cells from being killed, thus preventing the loss of insulin-producing cells. The loss of islet cells results in diabetes. This is just one of the roles played by DCs. Dendritic cells are ubiquitous in tissues and organs. They seem to manage immune responses by their ability to not only direct T cells, but also sample the local environment through their ability to endocytose (eat) debris exactly like macrophages.

DC Interactions with T cells that Affect the Anti-cancer T cell Response

There are two types of interactions between DCs and T cells: DCs instructing T cells and T cells instructing DCs. In a sense this is a two-way street. Imagine traffic flowing in one direction: DCs presenting cancer peptides to T cells that have a receptor for it. The T cells divide wildly to overwhelm the cancer and kill it. The T cell division is intense and all of a sudden a mutation occurs in the T cell receptor (TCR) gene so that a certain number of T cells begin to attack normal cells. These normal cells usually are the ones the cancer developed from. Nonetheless, they are being killed at an increasing rate. In the other direction, DCs are constantly producing Tregs. These lanes merge and the Tregs come in and shut down every T cell response in the vicinity. The cancer cells sit there for a while, but eventually they realize they are no longer being killed and go on their merry way to overcome the host.

This is an oversimplification because it leaves out the how and why, but you get the idea. The dualistic nature of the relationship between the DC and the T cell during the anti-cancer T cell response is crucial to the outcome of a cancer diagnosis. We will examine the specific mechanisms involved in this dualistic relationship later, but now let's focus on the physical part of this interaction. Dendritic cells and T cells communicate in two ways: through the secretion of lymphokines and through direct cell-to-cell contact. Both kinds of cells have the capability to secrete bio-active lymphokines and do so when instructed. The cell-to-cell contact takes place through a series of cell surface molecules that can signal the cell to perform a certain task. This refers to both cells, as will be clear as we delve into specific mechanisms these cells use to communicate with. This is at the heart of the next generation of cancer treatments. Controlling these events may mean the difference between life and death of a cancer patient.

T cells that respond to cancer cells elaborate a plethora of lymphokines (Hung et al. 1998). These small bio-active proteins find their receptor on the appropriate cell. Once that cell receives the lymphokines, it is internalized. This event begins a cascade of intercellular activity that results in a specific action that cell must perform. Monocytes receiving IL-4 and GMCF lymphokines differentiate into DCs. This differentiation occurs in two stages. If the signal is limited, the resulting DCs are of the immature type. Immature DCs frequently induce T cells to become Tregs. If this lymphokine signal is constant, mature DCs develop. The mature DCs, unlike the immature ones, can enter a tissue where there is an ongoing immune response, sample or absorb the proteins, and then find their way to the nearest lymph node. Once they reach this important immunological organ, the DCs begin to not only present the peptides from the tumor site to T cells, but also instruct T cells where to go and what to do there.

IL-12, IL-10, and gamma-interferon (INF-g) are three important lymphokines. These proteins, two of which (IL-12 and IL-10) are secreted by both DC and T cells, are pivotal during the anti-cancer T cell response. INF-g is mainly secreted by T cells, but under special circumstances DCs can also secrete it. IL-12 activates type-1 helper T cells and leads to a productive anti-cancer T cell response, which includes induction of CD8 killer T cells. IL-10, however, induces suppressive responses and has been shown to directly induce Tregs with another lymphokine called transforming growth factor beta, or TGF-B. INF-g secreted by CD8 T cells during a T cell immune response has the ability to increase expression of MHC class II proteins on DCs. This is important because it increases the DC's ability to present cancer peptides to responding T cells. Once CD4 T cells begin to respond to these peptides, they secrete more INF-g and IL-12, which can accelerate the response (Cohen 1990). However, there is a caveat to this situation. Tregs can be cancer-peptide–specific as well (Wang et al. 2005). As a matter of fact, some of these Tregs were at one time anti-cancer–responding T cells (Liu X et al. 2009). They turned into Tregs because they were converted by DCs. Tregs secrete IL-10 (Jonuleit et al. 2000) and once DCs stimulate them under these conditions, they begin to secrete massive amounts of this lymphokine. This causes DCs to secrete TGF-B and more Tregs are created (Marguti et al. 2009). So, if DCs enter a cancerous mass, and there are large numbers of Tregs present, there will be no productive anti-cancer response at that time. This is another very important aspect of the relationship between DCs and T cells that is crucial to control during immunotherapeutic treatment of cancer.

The pDC tends to immediately induce Tregs. It does this in a specialized manner by secreting an enzyme called indoleamine 2, 3-dioxygenase (IDO), which interferes with the ability of T cells to metabolize tryptophan. Tryptophan metabolism is essential for T cell function because, without it, all T cell-based activity ceases. IDO is a powerful immunosuppressant, but it can convert normal naïve T cells into classic T regulatory cells capable of secreting the immunosuppressive lymphokine IL-10 (Baban et al. 2009). IL-10 in combination with TGF-B can convert even more T cells into Tregs. Unfortunately, tissue DCs can also secrete IDO when the B7 on the DC surface is cross-linked by a Treg-expressed protein called cytotoxic T cell lymphocyte antigen-4 or CTLA-4. Once the B7 protein on the surface of DCs is cross-linked this way, IDO is released by DCs, which again has the same effect: the conversion of normal T cells into T regulatory cells. Fortunately, this IDO pathway is well known because when it breaks down tryptophan certain metabolites are produced, which are thought to actually induce the conversion of T cells to Tregs (Munn et al. 2004). This pathway has a pharmacological inhibitor, which will be discussed in the section on how to improve the ability of therapeutic DCs to induce a positive anti-cancer T cell response.

The next area to concentrate on is the ability of T cells to control DCs. Mostly this is accomplished by cell-to-cell contact using a series of ligands (keys) and receptors (locks). One of these receptors is called CD40, expressed by the DC. Its ligand CD154 (high) is expressed by activated T cells such as activated effector CD8 and CD4 helper T cells as well as platelets. Ignoring platelets and focusing on T cells, contact between CD154 on T cells and CD40 on DCs results in maturation of the DCs. This enables DCs to secrete the type-1 CD4 helper lymphokine IL-12 and present antigen at a much more efficient rate. It has also been shown that the Tregs begin to express the CD154 ligand after activation. However, the expression of this ligand is at minimal levels on Treg cells (Ma and Clark 2009). Perhaps the strategy here is for the Treg to only minimally mature the DCs, allowing them to continue to induce Tregs as immature DCs are able to. Despite that, the main purpose of this interaction is to amplify the anti-cancer T cell immune response. Another result of this interaction is that DCs induced into the immature (Treg-inducing) state may not stay at that stage for an extended period of time. Once exposed to CD154 (high) T cells, the DCs quickly get pushed into the mature DC phenotype. There, they can begin to induce the type of T cell response that invariably leads to the destruction of the cancer. This is called "licensing": activated CD8 killer T cells and helper T cells give DCs a license to do what they are supposed to do, which is to continually induce a productive immune response against the cancer (Hanig and Lutz 2008). This is why it is important that the DCs arrive at a cancer site where there are a dominating number of anti-cancer T cells. In this way they actually join the battle. However, if DCs arrive at a cancer site dominated with Tregs, the results are quite different.

Now let's talk about how Tregs control DC function in a way that is detrimental to the anti-cancer T cell response. Tregs express a gene called lymphocyte activation gene-3 (LAG-3 or CD223), a protein similar to the identifying CD4 protein found on normal helper T cells. This protein, once expressed, is a ligand. Its receptor is the MHC class II protein expressed by DCs and activated T cells. This interaction will probably cause many problems when DCs are used during anti-cancer immunotherapeutic treatments. Once DCs mature, there is an increase of certain cell surface proteins that enable the DCs to communicate with T cells in a positive manner. These are the receptors CD80 (B7.1) and 86 (B7.2). Their expression is an essential part of the maturation process that the DC must go through in order to sustain a productive anti-cancer T cell response. Often, Tregs encounter DCs arriving at a cancer site in an immature state because the lymphokine concentration that drew them there was not strong enough to mature them. The Tregs are drawn to these DCs like magnets. The LAG-3 ligand immediately connects with the MHC class II of the DC and something happens.

The interaction between LAG-3 and MHC class II induces the association of the cytoplasmic (inside of the cell) domain of MHC class II protein with other intercytoplasmic molecules called ITAM. Once this interaction occurs, a negative signal is sent to the DC. That signal tells the DC to lower its expression of CD80 and 86 to a level that makes it virtually impossible to communicate with T cells in a positive fashion (Liang et al. 2008). There are "innate" Tregs created in the thymus and there are the "adaptive" Tregs created from existing cancer-specific T cells (Bluestone and Abbas 2003). It turns out that both innate and adaptive Tregs can perform this type of DC inhibition, but the Tregs created from the anti-cancer T cell pool can do it even more efficiently. Published reports reveal that it takes only one-tenth as many cancer Tregs as it does innate Tregs to inhibit DC maturation. Unfortunately, once Tregs have basically de-differentiated the DCs, they have the ability to reprogram the DCs to go to the nearest draining lymph node (Liu Z et al. 2009). This is a problem because inside that lymph node T cells are being armed by antigen-presenting DCs to go into the cancer mass and destroy the tumor. Because of this Treg-DC conversion, a phalanx of Treg-altered DCs show up, ready to turn those productively produced anti-cancer T cells into more Tregs. The newly transformed anti-tumor Tregs go back into the cancer mass and stop any true anti-cancer T cells from doing their job. This, along with the LAG-3/MHC class II interaction, is possibly at the cause of the poor clinical results using DCs in what is called "cellular vaccines", but there is least one other cell-to-cell contact event that may very well be pivotal to the outcome of this scenario.

Use of DCs in Immunotherapeutic Treatment and Its Failure Due to Tregs

There is considerable evidence that T regulatory cells have controlled or interfered with anti-cancer immunotherapeutic clinical trials over the years. Tregs can exert control not only over other T cells, but on DCs as well. Immunotherapeutic treatment of melanoma is a good place to start because many of these treatments demonstrated measurable clinical responses (Rosenberg et al. 2009). However, very few of these treatments were unable to stop the progression of metastatic melanoma. For instance, tumor-infiltrating lymphocytes (TIL) isolated from a patient's tumor were expanded and infused back into the patient with the lymphokine IL-2. This therapy resulted in regression rates between 45 and 60 percent (Smith et al. 2008). The regression rate of this therapy was increased to above 70% by the use of a lymphoblation technique called total body irradiation. This treatment temporarily removes lymphocytes from the circulation and the bone marrow (Dudley et al. 2008). However, DCs, like many other myeloid cells, are radio-resistant, so their numbers remain intact. This means that Treg-"altered" DCs still have the ability to limit anti-cancer T cell responses. It was determined that the treatment removed Tregs that were interfering with the anti-melanoma T cell immune response. One additional problem with this entire immunotherapeutic approach is that Tregs are a component of the TIL, yet they were still used in the therapy. Fortunately, Tregs proliferate poorly *in vitro* so their numbers dwindled with each cycle of T cell division during the TIL expansion phase. Nevertheless, when these anti-melanoma T cells were returned to the patient, they came under the combined control of tumor-specific Tregs and Treg-altered DCs.

Vaccines based on DCs for use in immunotherapeutic treatment protocols have an inherent problem with T cell regulation. In fact, DCs and Tregs have a very complex relationship when it comes to the T cell response to cancer. However, DCs can actually control and manipulate Tregs. The presence of DCs loaded with tumor antigens may sometimes be an invitation for Tregs to get involved in the subsequent anti-tumor T cell response. For instance, a Phase I/II clinical trial was performed using DC-based therapy with or without low-dose IL-2. Sixteen patients were given DCs sensitized to autologous melanoma cell lysates and 18 patients were given DCs pulsed with melanoma-associated peptides representing six well-known MAA. One half of each group was given low-dose IL-2 and the other half was not. The results were two partial responses and three stable disease clinical responses in the nine patients receiving IL-2 and peptide-pulsed DC. There was one partial and one stable response in the other half of this group that did not get IL-2. Two out of eight patients developed stable disease in the melanoma cell lysate-pulsed DC group that did not get IL-2 and no clinical responses were reported in the corresponding group that did get low-dose IL-2. Statistically, there were no differences among the groups. This despite the fact that melanoma lysate-pulsed DCs were able to induce CD8 CTL *in vitro* (Redman et al. 2008). Clearly, something (Tregs) was preventing the development of a significant T cell immune response that would have led to the regression of these melanomas.

New Phase I/II trials were performed targeting the patients' DCs and their ability to present known melanoma-associated peptides gp100 and tyrosinase. This trial made use of the lymphokine GM-CSF, which can cause monocytes to mature into antigen-presenting DCs. The vehicle for its delivery was a DNA vaccine injected subcutaneously. This vaccine continually makes the protein, giving the body a continuous dose of this lymphokine. A total of 19 stage II/III metastatic melanoma patients were enrolled in the trial. Some of the patients were segregated into three groups of three patients each. Each group received an increasing dose of the DNA vaccine and a constant amount of peptides. The remaining patients received the highest dose of vaccine. The results indicated the treatment was safe with low toxicity at the vaccine injection site. Forty-two percent of the patients developed CD8 T cell responses measured *in vitro* to the peptides. There was no increase in CD8 activity corresponding to the increase in DNA vaccine concentration used. There was no noticeable change in clinical status in any of the patients (Perales et al. 2008). Lymphokine therapy directed specifically at DCs failed to induce cancer regression, despite measurable CD8 T cell responses *in vitro*. The likelihood is that Tregs *in vivo* were recruited as a result of DC activation and were able to inhibit these anti-melanoma CD8 T cells from attacking the tumor.

These are just a few examples of cases in which Tregs have interfered with or inhibited the effectiveness of immune-based cancer therapy protocols. In almost every case, as a result of the therapy, a clear induction of an anti-tumor T cell response could be demonstrated *in vitro*. In some instances *in vivo* detection of antibodies could be observed. However, none of these therapies were ultimately able to induce significant regression of the tumor. The examples above were selected because of their direct or indirect use of a patient's DCs in the therapeutic protocol examined. These DC-based therapies were formulated from years of accumulated knowledge suggesting that DCs played a pivotal role in the generation and maintenance of the anti-tumor T cell response. That evidence related to the DC's ability to efficiently present antigen, generate CD8 CTL, and secrete T helper type-1 (Th-1) CD4 T cell inducing lymphokines was initially derived from *in vitro* studies. Later, these findings were reproduced in mouse models. However, it is now being revealed that DCs play a dual role in controlling T cell tumor immunity. This has to do with the DC's ability to convert T cells and Tregs into each other's phenotypic characteristics.

Intercellular Events within the DC that Alter Their Function

Therapy using DCs has been largely ineffective although there is evidence that it can elicit an anti-cancer T cell response. There are issues surrounding this type of immunotherapy that have to be addressed for the therapy to be successful (Dauer et al. 2008). First of all, DCs can go into a cancer mass immunotherapeutically armed to induce an anti-cancer T cell response, but sometimes they induce even more anti-cancer suppression. Second, once immunotherapeutic DCs reach the cancer site, Tregs are immediately drawn to them. These two cells have an affinity for each other because of the T cell LAG-3/DC MHC class II interaction and the T cell PD-1/DC B7 interaction (below). Once this happens, the DCs that were supposed to induce an anti-cancer T cell response are at the mercy of Tregs, which dominate over them and their function. Tregs can actually alter the DCs, send them away from the cancer site, and tell them to produce more Tregs in the lymph nodes where anti-cancer T cells are developing. So this use of DCs as immunotherapeutic weapons is fraught with problems that scientists must address.

T cells can induce DCs to do something that is potentially dangerous to the body. Dendritic cells express a receptor called B-7. The ligand for the receptor is expressed by activated T cells (including Tregs) and is called PD-1. Once the PD-1 ligand on T cells attaches to the B-7 receptor on the DC, these receptors aggregate on the DC cell surface. The intercellular domains of the receptors begin to send signals to the DCs. This signaling occurs in conjunction with an adapter protein called DAP12 and protein kinases called syk and phospholipase C. These intercellular proteins become phosphorylated and the signals are created. These signals tell the DCs to secrete a lymphokine called IL-17, directed at CD4 T cells. CD4 T cells, which have the receptor for this lymphokine, attack "supposedly" normal tissues. These cells have the ability to break self-tolerance. If you remember the islet cell/steroid scenario then you know that this activity of Tregs to control anti–islet-specific T cells is a classic example of self-tolerance. The Th-17 T cells can break that tolerance and if they don't attack the islet cells themselves then the original anti-islet T cells will. The really strange thing about this is that a conversion takes place as far as T cells are concerned. Often when this happens a CD4 T cell is converted into a TH-17 T cell. More often than not, the T cell being converted is a Treg, which for all intents and purposes can be antigen-specific in nature

(Radhakrishnan et al. 2008a). In a sense, this activity can be viewed as a major problem, but actually it may not be.

Tregs come in two varieties: innate and adaptive. Adaptive Tregs are converted from the anti-cancer–specific T cell pool. If these cells are further converted into the TH-17 effector phenotype, they will have the ability to actually attack the cancer. This activity alters T cells that would normally suppress the anti-cancer T cell response. Now, these cells can attack the cancer and significantly contribute the anti-cancer T cell response. Unfortunately, the innate Tregs in the vicinity of this conversion also become TH-17 autoimmune effectors and can potentially cause problems. Because they are antigen non-specific or tissue organ–specific, they interrupt the non-cancer self-tolerance situations. This activity often leads to a plethora of autoimmune reactions in a cancer patient. Breaking self-tolerance, more often than not, makes the patient susceptible to infections. This (immunosuppression) is not an uncommon situation for a cancer patient. Signals that are induced by T cells and cause DCs to secrete this particular lymphokine can be inhibited by pharmacological agents. Once DCs are treated with these agents, the pathway toward releasing IL-17 can be interrupted. This will be covered in detail as we discuss ways to improve the action of DC-based anti-cancer therapies.

Dendritic cells that are CD8+ and CD205+ are specialized in their ability to induce regulatory T cells in the intestine. Secretion of IL-10 and TGF-B by DCs is essential in initiating this conversion process. These lymphokines are initiated by some of the toll-like receptor (TLR) proteins that these cells express. The TLR family is made up of eight receptors recognizing pathogen-related molecular patterns. Five of these receptors (TLR 1, 2, 4, 5, and 10) recognize bacterial products, while the remaining three (TLR 3, 7, and 9) see viral nucleic acids. Some of the natural ligands of these receptors are lipopolysaccharides and unmethylated CpG DNA motifs, which are free-floating DNA from disintegrated bacteria. There are quite a few TLRs expressed by DCs, but ones that seem to be involved with IL-10 and TGF-B are TLR-4, TLR-2 and, to some extent, TLR-7/TLR-8 (Lombardi et al. 2009). The ability of ligated TLRs to induce secretion of these two lymphokines by DCs is thought to involve intercellular pathways dependent on p38 MAPK induction (Groschel et al. 2008).

Improvement of DC-based Anti-cancer Therapy by Alteration of DC Intercellular Pathways and Treg/DC Cellular Contact

One of the interesting things about the complex relationship between DC and T cells is that there is a general understanding about the elements that control the response. It was earlier unthinkable to be able to change an immune cell's function, but now it is becoming a reality. Immune cells do what they do because they receive instructions, mostly from other cells. Sometimes, situations such as autoimmune reactions, sepsis (massive bacterial infection), or inflammatory reactions can induce multiple conflicting signals. This results in mistakes. The accumulation of knowledge concerning the function of and relationship between immune cells (DC and T cells) can help in removing cancer.

Dendritic cells can be improved so that they induce an effective, long-lasting anti-cancer T cell response. First, let's reflect on how DCs can be altered by contact with Tregs via the T cell PD-1/DC B-7 interaction. Intercellular signals occur within the DC that can cause them to secrete a lymphokine that converts T cells into autoimmune effector cells. Again, this is a problem because the autoimmune effectors can break immune tolerance. The converted T cells may be cancer-specific Tregs and can be added to the anti-cancer T cell army (Radhakrishnan et al. 2009). However, if this conversion happens to any other CD4 helper T cell, the results can be disastrous for the body. The signal received by the DC from T cells has to do with an adapter protein (DAP12) and the phosphorylation two kinases (syk and phospholipase C). This activity can be inhibited by an inhibitor of syk called piceatannol. Piceatannol effectively blocks this pathway and thus inhibits DCs from converting T cells into the TH-17 effector phenotype. Piceatannol treatment does somewhat inhibit antigen presentation by DCs, but if these cells are pre-conditioned, and will be used immunotherapeutically, then the impact of the treatment should be minimal (Radhakrishnan et al. 2008b).

A particular gene produces a kinase protein (serine/threonine kinase) and is expressed in a wide variety of immune cells, including DC. The protein is called hematopoietic progenitor kinase 1, or HPK1. It is a known regulator of T cell functions through its ability to inhibit intercellular pathways leading to T cell activation such as the NFkB pathway. However, expression of this gene in DCs was found to control their ability to become activated. It does not inhibit DC activation but limits it. Mice were bred that lacked this gene and DCs from these mice were at least 10 times as sensitive to activation as those from normal mice. The HPK1-negative DCs from these mice not only were able to secrete IL-12, IL-4, and IL-6 (T helper type-1 lymphokines) at a much higher rate and intensity, they could also eliminate established tumors in normal mice (Alzabin et al. 2009). This is an incredibly important discovery because for the first time DC activation could be amplified to make the cells truly function as mediators of a productive anti-cancer T cell response. This alteration of DC activation has not been tried in human DCs, but the techniques to accomplish the procedure are already in place.

One way to silence a functioning gene in cells is called RNAi, or interfering RNA. A complementary copy of RNA written by the gene is produced in the lab. This copy is introduced into the cell, where it binds to the gene-made RNA, which prevents the protein product of that gene from being expressed by that cell. In other words, protein production from that particular gene is effectively blocked. This is a selective way to temporarily silence a gene (Ross 2003). In this way DCs prepared for immunotherapy could have the HPK1 gene silenced prior to infusion into the cancer patient. The HPK1 gene-silenced DC can activate a productive anti-cancer T cell response before being influenced by Treg cells. Since the DCs now produce significantly higher levels of CD80 (B7.1) and 86 (B7.2), they should be capable of activating T helper type-1 T cells immediately.

Another pathway that leads to DCs producing IL-10 and TGF-B is through the TLRs. Activating DCs with ligands of these receptors and then interfering with the pathways associated with the p38 MAP kinases induces the production of T helper-1 lymphokines. Activation of DCs through the TLR is a surefire way for DCs to induce Tregs. However, when they are induced and the p38 pathway is inhibited, the DCs not only stop inducing Tregs, but also produce IL-12, IL-4, and INF-g. For INF-g, this is the special case where DCs secrete this lymphokine (Jarnicki et al. 2008). It is a special kind of alteration of DC function because it guarantees that the DC will begin to induce the kind of T cell response that is the starting point of a productive anti-cancer T cell response. Again, we are talking about DCs that have been prepared to be used in an immunotherapeutic clinical trial. These cells, having been removed from a cancer patient as monocytes and turned into cancer-peptide antigen-presenting DCs in the lab, are now stimulated by TLR agonist, treated with an inhibitor, and then returned to the patient as a treatment against the patient's cancer.

Both pDC and DC can under certain circumstances secrete the enzyme IDO. The enzyme generally inhibits T cell function by removing the amino acid tryptophan. The metabolites from the break-down of this T cell-dependent amino acid contribute the induction of Tregs. This induction involves not increasing Treg numbers by cell division, but converting Treg negative normal T cells into the phenotype (functional characteristics) of T regulatory cells. Experimentally, Tregs can be induced without the presence of IDO but in the presence of tryptophan metabolites made in the lab. A pharmacological inhibitor called 1-methyl-D-tryptophan (1MT) effectively blocks the activity of the IDO enzyme. Treatment of IDO-secreting DCs with this agent has shown that the cells no longer secrete active IDO, which stops the production of Treg-inducing tryptophan metabolites (Liu X et al. 2009a). The effects are temporary unless the inhibitor is given systematically. However, this may be enough time to prevent therapeutic DCs from making the active form of IDO so that a productive anti-cancer T cell response may occur right away without the induction of Tregs.

Cell-to-cell contacts between Tregs and DCs interfere with the ability of DCs to induce a productive anti-cancer response. The interaction of the Treg LAG-3 receptor with the MHC class II (ligand) on DC results in the inhibition of DC activation. The ability of a DC to become activated is essential for anti-cancer T cell response. The inhibition is due to DC intercellular events that occur as a result of this interaction. These events are associated with an ITAM-associated negative intercellular pathway. The pathway can be activated by other interactions that cause cross-linking of the MHC class II molecule. In the end, this activity blocks the ability of DCs to mature and express essential cell surface molecules, such as CD80 and 86, which are needed to induce a positive T cell response (Liang et al. 2008). In order to block the interaction between LAG-3 and MHC class II, it would be prudent to use antibodies against the class II molecule. However, that is impractical. For DCs to present cancer peptides, the MHC class II molecule must not be blocked. Instead, a series of inhibitors has been developed that would block ITAM and other related pathways, such as NF-kappa B, ERK, and p38. Manassantin A inhibits the latter pathway (Kim et al. 2009) and methyl-B-cyclodextrin inhibits the ITAM inhibitory pathway (Cifuentes et al. 2008). If immunotherapeutically prepared DCs are treated with these inhibitors prior to their infusion back into the patient, the DCs would be resistant to this type of Treg control.

One more cell-to-cell contact is the Treg inhibitory contact. In this case, antibody treatment of the DC is warranted. Tregs preferentially associate with DCs. They are exceptionally mobile and seem to find DCs right away. All matured T cells can move at an astonishing rate. This other aspect of cell-to-cell contact of Tregs controlling DCs has to do with cell surface proteins expressed by both cells. LFA-1 molecule expressed by Tregs and ICAM-1 expressed by DCs interact, resulting in the aggregation of Tregs and immature DCs and in the down-modulation of CD80 (B7.1) and 86 (B7.2), which prevents DC from activating tumor antigen–specific naïve CD4 T cells. The activity can result in tumor antigen-specific immune suppression and tolerance. Blocking ICAM-1 with a non–cross-linking antibody could thwart the efforts of Tregs to induce inhibition of therapeutic DC by down-modulation of CD80 and CD86. This interaction between the two cells leads to the initiation of other cell-to-cell contacts (Onishi et al. 2008).

Although the LFA-1 protein is involved with DC activation of naïve anti-cancer T cells, blocking the interaction between LFA-1 and ICAM-1 with an antibody would seem to hinder this activation. However, antibody treatments of cells like this are temporary because the antibody-coated ICAM-1 proteins are soon discarded by the DCs and replaced with new unblocked proteins. Since Tregs are at least 10 times as mobile as naïve T cells, they reach the immunotherapeutic DC first. Consequently, if ICAM-1 were blocked, the tumor Tregs would move on because they would be unable to attach themselves to the DCs. By the time the DCs shed and replaced the antibody-ICAM-1 proteins, normal naïve T cells would be close enough to become activated by the now fully functional therapeutic DCs. The DCs could also be induced to secrete IL-12, which attracts naïve CD 4 T helper type-1 T cells. Within the same immunotherapeutic DC preparation, there are elements that would resist Treg control and invite T cells that participate in a positive anti-cancer T cell response.

What you have just read may be at the heart of future anti-cancer treatment. Dendritic cells play an integral part in the anti-cancer T cell response, but they also control the entire T cell immune system in such a way as to make them invaluable in the immune system. If the DCs in our bodies disappeared, it would be catastrophic. We would suffer from a long list of immune diseases that would lead to death. Dendritic cells are special in that they are created from cells unlike them. Monocytes and macrophages have a special role in the maintenance of our bodies, but the DCs are the directors of the immune system. They do what their progenitors do, but they do it much better. They actually cannot eat as much as a macrophage, but whatever they eat is converted into T cell-recognizable peptides at 10 times the rate of the cells they were derived from. Dendritic cells can control the activities of T cells, yet T cells can control them. In the end, because of their versatility, they are the probably the most important element in the future treatment of cancer. Because DCs are so flexible, they are the perfect vehicle with which to fight cancer. Monocytes and macrophages can kill cancer cells. Macrophages and DCs do this by antibody-dependent cellular cytotoxicity. The immune system makes antibodies against the tumor. The antibodies find the tumor and adhere to its surface. Macrophages and DCs have a receptor called an Fc receptor. This receptor binds to the Fc portion of the antibody, which is the portion not attached to the tumor cell. Once the two cells are attached to one another, the macrophage/DCs bore a hole in the cell, killing it (te Velde and Figdor 1992). They perform this function in a limited way depending on the type of T cell response (TH-1 CD4 T cells induce tumor killing and TH-2 CD4 T cells activate B cells that produce antibodies). The ability of CD8 T cells to kill cancer cells dwarfs the killing function of the myeloid-lineage cells. However, when you couple this with the other things DCs do it is clear that DCs are more than special; they are invaluable in fighting cancer.

Since DCs are relatively easy to make in the laboratory and can rapidly present tumor antigens, their use in immunotherapeutic treatment protocols makes sense. The problems encountered with their use for this purpose are being understood by scientists. The way to approach these problems is to look within the DC itself. This has to do with the emerging field of translational immunology. Many immune cells do what they do because of intercellular signals and pathways that are activated by a wide variety of stimuli. These signals, generated by the interaction of DCs with not only T cells, but also soluble immune peptides (lymphokines) secreted by other immune cells, tell the cells what to do and for how long. The DCs themselves have the ability to signal other cells such as T cells by their response to some of these stimuli. Here we have a situation where both cells (DCs and T cells) are signaling each other to perform a specific task. The question scientists must answer is, which of these tasks will destroy a cancer?

Chapter 11

Regulatory T cells, Dendritic Cells and the Success of Immunotherapeutic Treatments for Cancer

T regulatory cells (Tregs) play an essential role in the induction of self-tolerance by suppressing autoimmune responses (Sakaguchi et al. 2001). Many of these responses are tissue-specific, yet some are related to the development of tumors and can be detrimental to the anti-tumor T cell response. Despite this, T regulatory cells can be converted from naïve or antigen-specific T cells in the context of a growing tumor (Drake et al. 2006). This is indicative of an important anti–self-tolerance mechanism. Even though this is not the only immunosuppressive T cell mechanism controlling the anti-tumor T cell response (Edwards and Evavold 2010), it probably represents one of the main reasons why many anti-tumor T cell responses are ineffective. The activities of Tregs during an anti-tumor T cell response may be crucial to the success or failure of such a response while the tumor is growing (El Andaloussi and Lesniak 2006). Both "innate" and "adaptive" Tregs play a role in the anti-tumor T cell response, but the adaptive Treg in this situation may be a pivotal element because of the T cell/Treg conversion phenomenon (Bluestone and Abbas 2003). The conversion of naïve or tumor-specific T cells into Tregs, and Tregs into antigen-specific effector T cells, is mediated by DCs (Curiel 2007). Dendritic cells not only mediate this conversion or reprogramming process, but also are essential to the expansion and maintenance of the Treg (Yamazaki et al. 2003). They can also themselves come under the control of Tregs, and thus alter their function in such a way as to benefit tumor growth (Liang et al. 2008).

Immunotherapeutic treatment protocols, which use DCs as cellular vaccines, have met with difficulty because of the ability of DCs (particularly immature cells) to convert T cells into Tregs. In addition, other immunotherapeutic treatments can induce tumor-related autoimmune reactions, which may be the result of Tregs being converted into autoimmune effector T cells (Radhakrishnan et al. 2008a). This bi-directional conversion or reprogramming of the Tregs by DCs is probably one of the most important findings directly related to the success or failure of immunotherapeutic treatments for cancer. The manipulation of this process during DC-based immunotherapy is essential to the recruitment and generation of a sustained anti-tumor T cell response. Temporary interruption of the conversion of naïve or tumor-specific T cells to Tregs would give the T cell immune system additional time to eliminate the tumor. Conversely, interruption of the conversion of Tregs to these autoimmune effector T cells would prevent tumor-induced autoimmune reactions from occurring because of the reduction of self-tolerance–inducing Tregs and the creation of proinflammatory anti-self Th-17 CD4 effector T cells (Juszczak 2009).

Regulatory T cells (CD4+, CD25+) are further defined by their expression of the Foxhead box P3 transcriptional repressor or FOXP3 gene. Stable expression of this gene by T cells typifies the regulatory phenotype of these cells (Walker et al. 2003). However, the FOXP3 gene is transiently expressed by activated T cells and hyporesponsive T cells cultured with TGF-B (Ziegler et al. 2007). Despite evidence suggesting that FOXP3 gene expression under some of these conditions does not lead to the Treg phenotype, there is possibly a "flux" state in which such T cells possess the phenotypic characteristics of both naïve and antigen-specific T cells and Tregs. This may explain situations where removal of T cells based on the Treg phenotype results in an increase instead of a decrease in tumor growth, which is often associated with Treg removal (Curiel 2007). In order to manipulate these conversion processes mediated by DCs, an in-depth understanding of the process is warranted. This could lead to the identification of crucial points in these processes where inhibitors of intercellular pathways, which lead to either Treg conversion event, can be manipulated to benefit the host anti-tumor T cell response. Similar strategies can be applied to Treg control of DC function. These inhibitors can be administered in a non-systematic manner by treating only immunotherapeutic DCs used in the procedure. This would lead to modulation of the conversion events to allow the anti-tumor T cell response to proceed without Treg inhibition or the development of tumor-related autoimmune responses.

Conversion of Tregs into Autoimmune Effector T cells

The first of these two conversion events induced by DCs, which results in Tregs being converted into effector T cells, is just beginning to be understood. Studies have shown how DCs, after activation with antibodies to B7-DC/PD-L2 molecules, can effectively convert Tregs into autoimmune effector T cells. FOXP3 gene expression is down-regulated, IL-10 and TGF-B secretion stops, and the cells are no longer able to suppress T cell responses. This interaction between activated DCs and Tregs was found to be dependent on IL-6 secretion and was antigen-specific. The converted effector cells secreted INF-G, TNF-A, and IL-17 and were thought to be TH-17 helper effector T cells capable of breaking self-tolerance and inducing autoimmune disease in an antigen-specific manner. Because this activity does not result in widespread autoimmunity but induces an anti-tumor T cell response, these autoimmune effectors may be attacking neo-antigens (Radhakrishnan et al. 2008a). This observation could have wide-reaching implications because there are tumor-associated Tregs that are tumor antigen–specific by virtue of the T cell receptor. If these T cells, through a similar DC-dependent process, can be converted into true anti-tumor effector T cells during immunotherapeutic treatment, it would be of tremendous value to the success of such a procedure. Of course, this would have to be a transient conversion because the loss of tumor antigen-specific Tregs could lead to tumor-associated autoimmune reactions.

The Th-17 CD4 T cell subset is a special type of CD4 T cell (Oukka 2008). Apart from being able to break self-tolerance, which prevents autoimmune reactions from destroying tissues and organs, Th-17 CD4 T cells are potent inducers of tissue inflammation. If the lymphokine IL-17 secreted by these cells is not regulated, IL-17 in conjunction with other lymphokines can initiate organ-specific autoimmunity (Korn et al. 2009). The generation by DCs of Th-17 CD4 T cells from Tregs has several interesting aspects. First, Th-17 T cells are the essential CD4 T cells that mediate autoimmunity and inflammation. Inflammation induced by the lymphokines (IL-17A, IL-17f, and IL-22) secreted by these T cells stimulates fibroblasts (a form of skin cell), endothelial cells (cells that line blood vessels), epithelial cells (skin cells), and macrophages to produce bioactive small proteins (chemokines and cytokines) to recruit polymorphonuclear leukocytes (PMN) to sites of inflammation. The PMN secrete a host of bioactive chemicals that actually induce the inflammatory state (Cassatella 2003). On the face of it, this type of situation in a tumor mass would not be considered a good one because it can do two things. First, through Th-17 Tregs, it induces macrophages into an "alternative activation" phenotype. This type of macrophage secretes anti-inflammatory lymphokines and cytokines that not only generally induce T cell suppression, promote tumor growth, and result in tissue remodeling, but also induce the formation of Tregs from naïve T cells (Tiemessen et al. 2007). The appearance of these macrophages at a tumor site contributes to the diminution of the T cell response because these cells have reduced levels of MHC class II, B-7 (CD80), and B-7.2 CD86. That means they are unable to activate T cells by presenting tumor antigens

(Taams et al. 2005). Second, through DCs, the PMN bioactive chemicals can bring about the same result.

It has been demonstrated in mouse models that the Th-17 CD4 T cell itself is responsible for directly attacking the tumor and removing large established mouse melanoma tumors (Pavelko et al. 2008). The DC conversion does not lead to widespread autoimmunity; instead, it results in the destruction of the tumor. That says something about what is really happening as a result of the Treg to Th-17 conversion. This is a homeostatic balancing act. As Tregs are being converted into Th-17 CD4 effector T cells, the Tregs are being replaced by the two mechanisms mentioned above. Since many of the Tregs that have been converted into Th-17 effector cells were tumor-specific Tregs (possesing anti-tumor T cell receptor) that were previously anti-tumor effectors themselves (Liu X et al. 2009), before conversion (by DC) into Tregs, an anti-tumor T cell response occurs. This time, however, the anti-tumor effector cells are of the Th-17 subset and are secreting IL-17. The system depletes one type of cell (Treg), then replaces it while producing anti-tumor Th-17 CD4 effector T cells, which attack the tumor. That is why cancer-induced autoimmunity is not induced as one would expect from the depletion of Tregs, and consequently the tumor becomes smaller because of the generation of these newly formed tumor-specific Th-17 CD4 effector T cells. Understanding this pathway in the context of a growing tumor is important to the development of new anti-cancer immunotherapeutic treatments. Although Tregs seem to interfere with immunotherapeutic treatment protocols, they are an essential component of the system and have to be redirected rather than removed. The fact is that when these cells are removed in a systemic manner (anti-Treg antibody infusion), the consequences can be deadly for the patient, who would suffer from inflammatory or autoimmune toxicity (von Euw et al. 2009).

Conversion of Normal or Antigen-specific T cells into Tregs

The second conversion event (naïve/antigen-specific T cell to Treg) has been the subject of much study. The conversion into Tregs is important not only as far as cancer is concerned (Shimizu et al. 1999), but also in the control of organ-specific autoimmunity (Mottet et al. 2003). Dendritic cell control of this conversion event is well documented in tumor model systems and organ-specific autoimmune models. In tumor model systems, the tumor itself can control the ability of DCs to convert naïve or tumor-specific T cells into Tregs. This occurs either when the tumor inhibits DC maturation (Gabrilovich et al. 1996), which allows immature DCs to induce Tregs, or by the induction of TGF-B secretion by DC exposed to the tumor (Dumitriu et al. 2009). Moreover, the tumor itself has been shown to secrete

TGF-B directly, converting CD4+, CD25- T cells into Tregs (Liu et al. 2007). In addition, anti-tumor T cells can be converted into Tregs by this process. The presence of Tregs in tumor lesions (Jandus et al. 2008) is associated with the suppression of the anti-tumor T cell response (Grauer et al. 2007). Their presence not only in tumor lesions, but also in the draining lymph nodes (Viguier et al. 2004) is an indication of the type of control Tregs exert over the anti-tumor T cell response.

The conversion event from either naïve T cell or antigen-specific T cell to Treg is unique as far as T cells are concerned. Regardless of whether it is induced by DCs or tumor cells themselves, the conversion does not require cell division by the T cells getting converted (Walker et al. 2005). Rather, the T cell is transformed into this phenotype by the up-regulation of genes within the T cell that are normally silent. One such gene is the FOXP3 (Fontenot and Rudensky 2005). This gene can be transiently expressed in activated T cells without those cells becoming Tregs. Therefore, just the expression of FOXP3 in T cells is not an indicator of the Treg phenotype (Pillai et al. 2007). The phenotype itself is a functional definition. All adaptive (converted) Tregs are not the same in their specificities and modes of action when it comes to inhibiting other T cells from responding. This is a consequence of the conversion process. The conversion process itself is varied in its source and intensity. This results in the creation of various types of Tregs. For example, a naïve T cell converted into a Treg will behave according to the type of conversion signal, such as lymphokines (IL-10, TGF-B, IL-35, or combinations), direct cell-to-cell contact with DC, or cell-to-cell contact with a tumor cell. Each of these conversion-inducing methods creates a specific type of Treg.

Each type of converted Tregs has many ways to thwart the efforts of antigen-specific T cells to respond. First, they secrete the lymphokines mentioned above, which not only have the potential to convert these responding T cells into Tregs, but also can inhibit the responding T cells because they can retard cell division. The elaboration of these lymphokines by converted Tregs does not affect them because they do not need to proliferate to function, but antigen-specific responding T cells must divide to function properly. Tregs that use lymphokine secretion as a weapon against antigen-specific responding T cells were often converted from naïve T cells by exposure to massive amounts of the lymphokines.

Second, converted Tregs control antigen-specific responding T cells by the method used by CD8 effector T cells (controlled by Tregs) and NK cells. This is called cytolysis or cell killing. Briefly, once a CD8 killer cell or an NK cell is in contact with its target cell, the perforin enzyme bores a hole in the target cell and the contents of that cell leak out, which kills it. It was discovered that CD4 T cells had this function and now it is clear that Tregs also have it. Therefore, a Treg can not only inhibit an antigen-specific

T cell from responding, but also kill it before it reaches its target. This type of converted Treg was likely created by direct cell-to-cell contact with an activated DC.

A slightly controversial method Tregs use to inhibit antigen-specific responding effector cells is metabolic disruption. The first aspect of this function is basic. The idea is that the Treg, because it expresses massive amounts of CD25 (IL-2) receptor, has the ability to absorb IL-2 needed by antigen-specific T cells to not only survive, but proliferate in response to antigen. It is controversial because there is conflicting data on just how effective the inhibition of T cell function is. Nevertheless, there is a logic behind it. The next metabolic disruption pathway has been proven. There is a metabolic inhibitor called cyclic AMP (cAMP) and Tregs possess it but it does not inhibit their metabolism because it is in messenger form. The Tregs have the ability to inject the cAMP message into CD8 effectors through membrane gap junctions. Once inside the cell, the message is read, cAMP is created, and the metabolism of the CD8 T cell is completely disrupted. The result is a non-functioning effector T cell (Sojka et al. 2008). This fact is being exploited by inhibiting cAMP production in Tregs (Bopp et al. 2010). This type of control Tregs have over antigen-specific effector T cells seems to be a generalized function not associated with the mode of conversion.

The way T cells get converted into Tregs may dictate the way Tregs inhibit antigen-specific T cell responses. This plasticity in the functional behavior of converted Tregs is based on the conditions under which the Treg was created. These examples were based on a naïve T cell being transformed into Treg, but CD4 antigen-specific T cell responders and effectors are converted in exactly the same way. However, because Tregs are specific for a particular cell or tissue, they are limited to those areas. In a tumor mass, the antigen-specific Tregs not only protect the tumor from attack by tumor-specific effector T cells, but also begin to influence DCs, which are part of the anti-tumor T cell response.

Treg Control of DC Function and DC-based Anti-cancer Immunotherapy

Tregs can not only control other T cells, but also influence DCs, which home to lymph nodes after being created at the site of the tumor by activated T cells (Bayry et al. 2007). Tregs do this by inhibiting DC activation (Liang et al. 2008), maturation (Onishi et al. 2008) and function (Andre et al. 2009). Immunotherapeutic trials using tumor-associated peptide-pulsed DC have demonstrated transient increases in Tregs immediately following the vaccination, which return to pre-vaccination levels. This increase, which is presumably a response to new anti-tumor T cell activity induced by the

DC vaccine and the ability of DC to drive Treg cell division (Yamazaki 2003), may have been responsible for the success rate of less than 40 percent (4 of 12 patients; Lepisto et al. 2008). An overview of DC vaccinations has revealed that there are some favorable clinical outcomes, but only in a minority of patients (Aarntzen et al. 2008). This control that Tregs have over the function of DCs is increased when the Tregs are derived from a growing tumor. The number of tumor Tregs needed to down-regulate CD80 or CD86 expression is one-tenth the number of natural Tregs needed to perform this DC-inhibiting function. This may be a mechanism tumor Tregs use to disarm tumor antigen-presenting DCs (Liu et al. 2009b). In addition, Tregs can have a profound effect on maturation changes in DC that lead them to migrate directly to lymph nodes, where both Tregs and anti-tumor effector cells are generated (Bayry et al. 2007). Once in the lymph nodes, regardless of the fact that they are presenting pulsed tumor antigens, therapeutic DCs are susceptible to Treg control. In fact, removal of Tregs prior to the use of a tumor RNA-transfected DC vaccine greatly increases the effectiveness of the vaccine in renal cell carcinoma, melanoma, and glioma patients (Dannull et al. 2005; Grauer et al. 2008).

This control of DCs by Tregs is complex and almost immediate. The relationship between Tregs and DCs is a dualistic one. It would have to be, if we look at the big picture of immunity. These two cells are the only immune cells responsible for distinguishing between responses that are anti–non-self (bacteria, viruses, and tumors) and anti-self (autoimmune). In some instances Tregs control DCs and in other situations DCs control Tregs.

How can controlling the relationship between these two powerful immune cells increase the effectiveness of anti-cancer immunotherapeutic treatments? In the rest of this chapter, we outline how and why Tregs interfere with this promising new anti-cancer therapy. Because immunotherapeutic treatment of cancer can potentially remove a cancer forever, is relatively painless, does not require surgery, and has few side effects, it is an important area of research that must succeed. Anti-cancer treatments of the future will depend heavily on the knowledge gained by the research in this field. The key to success lies in understanding the relationships between Tregs and DCs and applying that knowledge to the design of improved immunotherapeutic treatments.

This section is probably the key to understanding the failure of modern anti-cancer immune cell–based treatments (immunotherapy) to consistently remove the cancer. Tregs interfere with or inhibit outright the ability of these treatments to slow tumor growth or have any effect at all. In early studies, tumor-bearing mice were treated with a form of immunotherapy called "adoptive transfer", where T cells were removed from normal mice of the same strain, immunized against the tumor, and then returned to the tumor-bearing mice. The initial observations were to test the T cells' ability to kill the tumor in laboratory experiments. These were called cytotoxicity assays. They were positive, but the real experiment was to determine which group of mice developed advanced tumors. Mice that received the immunized T cells did not develop advanced tumors. In humans this is a problem, where patients can be treated with these immunotherapeutic procedures, but the tumor remains; and the only evidence the procedure was even done is the ability of the patient's T cells to respond to the tumor in the lab.

An anti-cancer immunotherapeutic treatment is developed using, let's say, tumor antigen–pulsed DCs that had been genetically engineered to produce the lymphokine INF-g. In the laboratory and in mice, this procedure results in the generation of an anti-tumor T cell response that has been shown to reduce, if not remove, tumors from mice. The logic behind this is sound because the DCs present tumor antigens to naïve tumor-specific T cells. The INF-g secreted by the DCs induces MHC class II expression on the tumor and supports the generation of Th-1 CD4 T cell. The increased expression of MHC class II proteins by the tumor accelerates the response by stimulating more anti-tumor Th-1 CD4 T cells. The activation of these cells causes them to secrete IL-2, which matures tumor-specific CD8 killer T cells, causing them to attack the tumor. When this exact immunotherapeutic protocol is reproduced in human clinical trials, it fails, with only minor responses and the occasional reduction in tumor mass. When the results are compared to the mouse studies, this trial is a failure. The only evidence of effect of the treatment is that the majority of the patient's anti-tumor T cells have become activated. This observation is telling. The fact that the procedure did induce a T cell response to the tumor is established, but does that response not slow down or stop tumor growth? The answer is that the Tregs in the patient's body take control of the immunotherapeutic DCs. Tregs control DC activity mainly through two routes. These two routes can be used either separately or together. The first is cell-to-cell contact and the second is through soluble mediators such as lymphokines released by the Treg that affect the DC.

The first cell-to-cell contact mechanism Tregs use to control DCs has to do with a protein ligand on the surface of activated T cells and Tregs. It is called leukocyte function-associated antigen-1 or LFA-1. Its receptor is a protein on the surface of the DC called ICAM-1 (Onishi et al. 2008). This is an adhesion protein that allows the cell to adhere to tissues. Another aspect of this interaction is the mobility of the Treg cell compared to that of naïve T cells. A Treg is actually an activated T cell and moves rapidly like an amoeba. It has an affinity for DCs. Once the two proteins from both cells make contact it is as if they are glued together. Only after the Treg is satisfied that the DC is altered will it let go. The main point here is that the

Treg always finds the DCs first. Naïve tumor-specific T cells do not have the mobility to win the race, and they do not express LFA-1, which would allow them to adhere to the DC. So Tregs physically block naïve T cells from reaching the DCs. On top of that, once the Tregs are attached, they down-regulate two important DC cell surface proteins needed by naïve T cells to respond to the tumor antigens the DC presents. Those are CD80 (B7.1) and CD86 (B7.2). This interaction is the initial one between the Treg and the DC. Now you can see what happens to immunotherapeutic DCs that are infused into a cancer patient. They are immediately accosted by Tregs.

The next cell-to-cell contact method Tregs use to manipulate DCs is to send signals to the DCs to begin a conversion process. Activated T cells, as well as Tregs, express a receptor called PD-1. The ligand for this receptor is the B7 protein expressed by the DCs. Interaction between these two proteins can cause DCs to convert naïve T cells and Tregs into Th-17 CD4 autoimmune effector cells. Once the B7 molecules are aggregated on the DC surface by the PD-1 proteins of the T cells, a signal is produced just inside the DC cell membrane that induces phosphorylation. This step causes the aggregation of three proteins on the DC cell surface: B7 (CD80), B7.2 (CD86) and TREM-2. These three are thought to lead to the DC conversion process. This was discussed previously in the context of a somewhat closed tumor mass situation where tumor-specific Tregs were converted into Th-17 effector cells, which could then attack the tumor. However, there is more to this situation: initially Tregs induce DCs to convert Tregs into effectors, but naïve antigen non-specific T cells are also converted and they can do serious damage to other normal tissues and organs. The impact depends on the location of the reaction. In a solid mass, Th-17-converted Tregs may kill the tumor because there are few naïve T cells in the mass. However, if the tumor is diffuse or capillary-rich, there are a large number of naïve T cells at the site of this conversion, and all may be converted. The patient would surely develop an autoimmune reaction. This is called autoimmune toxicity.

Another cell-to-cell contact method Tregs use to change DC involves another Treg-expressed protein derived from a gene called lymphocyte activation gene-3 (LAG-3) and related to the CD4 proteins. It is a ligand and its receptor, like the CD4 protein, is the MHC class II protein expressed by the DC. When these two proteins bind, they make the MHC class II protein unavailable to naïve anti-tumor T cells. However, that is not the biggest problem. This interaction triggers intercellular changes within the DC that retard its ability to mature. This happens in a similar manner where the aggregation of the MHC class II proteins on the cell surface sets off a series of intercellular activation events that lead to what is called an ITAM-mediated inhibitory signaling pathway. Once this pathway is activated inside the DC, it inhibits DC activation, and DC maturation is halted. The interaction de-differentiates the cells, a reverse mutation. Once this happens, the DCs are in an immature state and induce Treg formation. The affected DCs no longer secrete IL-12 needed for an anti-tumor T cell response. Therefore, this interaction between the LAG-3 ligand on the Treg and the MHC class II receptor on the DC inhibits the activation of naïve anti-tumor T cells.

The final type of Treg control of DCs is the most deadly and, in some sense, the most effective way Tregs take control of the DC vaccine treatment of cancer. Let's closely examine the two cells involved. The first is the DC that has been prepared outside the body to elicit an anti-tumor T cell response. It does this by presenting tumor antigen to either naïve or activated innate or adaptive anti-tumor T cells. The DC used is pulsed (incubated) with either whole tumor cell or solubilized tumor cell membrane. It takes in these proteins and begins the process of antigen presentation, which involves linking tumor-derived peptides to MHC I and II molecules destined to be expressed on the cell surface and made available to the tumor-reactive T cell. With that in mind let's look at the Treg. Again there is an innate Treg, which is predestined to protect the tumor it perceives as a normal cell, and there is the adaptive Treg that was an anti-tumor T cell (CD4 or CD8) before it was converted into a FOXP3 positive Treg. Both CD4 and CD8 Treg (adaptive) still possess the killing machinery of the anti-tumor cytotoxic T cells they once were. Both can kill tumor cells via the recognition of self (MHC I or II) and tumor peptide on the cell surface. Therefore, when the prepared immunotherapeutic DCs are re-introduced into the patient, they find their way to the draining lymph node near the tumor site. Once there, these adaptive Tregs instantly kill the DCs through the classic CTL perforin method. This finding explains why DC vaccine immunotherapeutic trials fail and the DCs used for these treatments seem to disappear (Boissonnas et al. 2010).

Tregs control DCs using soluble mediators, namely lymphokines. This involves some cell-to-cell contact. Tregs have the ability to secrete copious amounts of one lymphokine (IL-10) and one cytokine (TGF-B). Both of these proteins alter DC function. Once DCs are exposed to them, they tend to secrete them, an autocrine response (Corinti et al. 2001). The two proteins can by themselves convert T cells into Tregs. However, both of them inhibit tumor antigen–specific T cell proliferation. Simply put, immunotherapeutic DCs arrive at the tumor site, are trapped by Tregs, and are bathed in a lymphokine mixture, which causes them to secrete the same lymphokines that inhibit T cell proliferation and induce the conversion of T cells into even more Tregs.

This scenario and the various cell-to-cell contact methods used by Tregs to control DCs probably explain why anti-cancer immunotherapies using antigen-presenting DC fail to demonstrate clinical tumor regression. However, now that these various mechanisms are understood, the immunotherapeutic protocols can be modified to neutralize many of the methods Tregs use to control immunotherapeutic DCs.

Combined Effects of T cell Conversion and Treg Control of DCs on Immunotherapeutic Treatment of Cancer

The two T cell conversions and Treg influence on DCs represent fundamental challenges to the effectiveness of DC-based cancer immunotherapy. This immunotherapy traditionally has been used to induce anti-tumor T cell responses based on the heightened ability of DC to present tumor antigens (Porgador et al. 1996) and has not taken into consideration the homeostatic relationship these cells have with the immune system (Merad and Manz 2009). Nor has it taken into account the obstacles DC vaccines create to the generation of a cancer-regressive T cell response (Kalinski et al. 2009). However, in addition to their role of facilitating an anti-tumor T cell response, DCs used for immunotherapy must be able to control T cell/Treg conversion events and resist Treg influences that impede their function for this type of immunotherapy to be truly successful. The objective response rates of these therapies for the treatment of melanoma are less than 10 percent (Schadendorf et al. 2006). In other studies, with late stage melanoma, objective response rates were 10–15 percent. However, T cell responses were detected in 60 percent of patients with regional lymph node metastasis and 30 percent of patients with metastatic disease (Aarntzen 2008). Such evidence supports the concept that DC-based therapy should work but does not, because of the influence of Tregs. This type of immunotherapy should result in the regression of tumors and possibly can if DCs can be instructed or restricted to induce only an anti-tumor T cell response while they control Treg function instead of Tregs controlling their function.

This section combines the total effects of three aspects of the relationship between Tregs and DCs and the difficulties that presents to the successful application of immunotherapeutic treatments for cancer. This is a difficult problem. There are no easy answers because these two cells interact with each other in unique ways. The problem is really a numbers game, but let's look at the easy solutions, which potentially are dangerous for the patient. It would be nice, for example, to remove or inhibit the Tregs of a patient prior to DC-based immunotherapeutic treatment, but over and over again this has proven to be a slippery slope. If Tregs are removed or inhibited there is the induction of autoimmune toxicity, which can be detrimental to the heath of the patient (von Euw et al. 2009). On the other hand, you cannot infuse massive amounts of tumor antigen-presenting

DCs into these patients either. The DCs, which then outnumber the Tregs, begin to generate an excess of anti-tumor T cells that eventually affect the DCs. The newly generated anti-tumor T cells begin to alter the DCs and do two other things. First, therapeutic DCs start converting T cells (both antigen-specific and non-specific) into autoimmune effectors. These effectors begin to break down self-tolerance situations all over the body and this results in the development of autoimmune toxicity. Second, over time the presence of so many anti-tumor T cells induces the DCs to convert both naïve and tumor-specific T cells into Tregs.

As we said, this is to a certain extent a numbers game. If the immunotherapeutic DCs are in the minority, there's a problem. Superior numbers of Tregs overwhelm the DCs that were prepared so carefully to induce an anti-tumor T cell response. These immunotherapeutic DCs become what amounts to fodder for the Tregs. The Tregs redirect (or kill) the DCs to convert more T cells into Tregs, leaving behind the useless anti-tumor effector T cells measured in the lab. This is the primary reason why immunotherapy based on DC biology is doomed to failure without intervention, although this type of therapy should work. However, it can work if scientists understand the underlying biological processes both the cells use to regulate the anti-cancer T cell immune response. That is beginning to happen because of advances in translational immunology. The next section will show how some of the intercellular pathways can be altered within both DCs and Tregs to increase the effectiveness of anti-cancer immunotherapeutic treatments.

Role of DC Intercellular Activation Pathways in Anti-cancer T cell Immune Response and How Blocking These Pathways Can Alter DC Behavior and Improve Immunotherapeutic Treatments

Treg to autoimmune effector T cell conversion and its dependence on DC intercellular activation pathways

Information is accumulating on intercellular pathways used to control not only phenotypic changes in DCs (Chen et al. 2008b) but also maturation (Liang et al. 2008) and function (Sun Y et al. 2009b). With this knowledge it should be possible to alter immunotherapeutic DCs to allow them to induce a productive anti-tumor T cell response. The previous conversion event that induces Tregs to become autoimmune effectors (Radhakrishnan et al. 2008a) is initiated by the cross-linking of the B7-DC ligand on DC. Cross-linking of the B7 ligand by antibodies or presumably the natural receptor PD-1 on activated T cells induces specific intercellular events. The events involve the phosphorylation of the adapter protein DAP12 and the protein kinases syk and phospholipase C. This activation pathway leads to increased antigen presentation by the DC. Blocking this pathway by treating B7-DC cross-linked DC with the syk inhibitor piceatannol results in a slight inhibition of antigen presentation and induction of protective immunity. But the ability of these cells to convert Tregs into autoimmune effector T cells is completely blocked. Blocking the conversion of tissue-specific Tregs into these Th-17 autoimmune effectors during immunotherapy using DCs to induce an anti-tumor T cell response would lessen the occurrence of autoimmune reactions not related to the response against the tumor. Therefore, in situations where there are few Tregs at the tumor site, inhibition of the syk pathway would be prudent. Since the function of reprogramming Tregs to be autoimmune effectors is an antigen-specific process, blocking this pathway should be considered important for immunotherapeutic control of the Treg to effector T cell conversion process (Radhakrishnan et al. 2008b).

DC intercellular activation pathway involvement with conversion of normal or antigen-specific T cells to Tregs

Blocking the conversion of naïve or anti-tumor T cells into Tregs has an obvious importance. This too is a function mediated by DC. Innate and adaptive Tregs can control not only the autoimmune activities of anti-tumor T cells, but also the anti-tumor T cells themselves. The adaptive or converted Tregs are special because if they have been reprogrammed from the anti-tumor T cell pool they retain their specificity for the tumor although they lose the ability to attack it (Wang et al. 2005). The retention of specificity keeps Tregs close to the tumor site so that they can stop newly generated anti-tumor T cells from erroneously attacking normal cells. However, many times these new tumor-specific Tregs cannot differentiate between auto-reactive T cells and anti-tumor T cells, so both are inhibited. This leads to tumor progression. Anti-tumor T cells are more quickly induced to become Tregs by the presence of tumor-derived non-self antigens in contrast to viral non-self antigens. This conversion occurs early in tumor development and at the transcriptional level (Getnet et al. 2009). Once the conversion is complete, both genes for FOX3 and CD25 are up-regulated.

Unfortunately, DCs are involved in this process as well. The regulation of these Treg genes is a direct result of the activation "state" of the macrophage or DC (Hanig and Lutz 2008). Dendritic cells that are CD8+ and CD205+ are specialized in their ability to induce regulatory T cells in the intestine (Yamazaki et al. 2008). Secretion of IL-10 and TGF-B by DCs is essential in initiating this conversion process (Perez et al. 2008). These lymphokines, particularly IL-10, are secreted by DCs in response to activation by some of the toll-like receptor (TLR) proteins that the

DCs express (Samarasinghe et al. 2006). The TLR family is made up of receptors that recognize pathogen-related molecular patterns and there are 10 of them. Five of these 8 receptors (TLR 1, 2, 4, 5, and 10) recognize bacterial products, while the remaining three (TLR 3, 7, and 9) see viral nucleic acids (Lombardi et al. 2009). Some of the natural ligands of these receptors are lipopolysaccharides (bacterial cell membrane components) and unmethylated CpG DNA motifs (free-floating bacterial or viral DNA). Dendritic cells express several TLRs, but the ones that seem to be involved with IL-10 and TGF-B are TLR-4 (Higgins et al. 2003), TLR-2 (Ghosh et al. 2006), and to some extent TLR-7/TLR-8 (Lombardi et al. 2009). The ability of ligated TLRs to induce DCs to secrete the two lymphokines is thought to involve intercellular pathways dependent on p38 MAPK induction. Once this pathway is blocked by the inhibition of p38, the TLR-activated DCs not only cease to convert T cells into Tregs, but also secrete Th-1 CD4 T cell-inducing lymphokines, such as IL-12, IL-4, and INF-g. Blocking the p38 pathway also augments the anti-tumor T cell response, presumably by the activation of TH-1 CD4 T cells (Aicher et al. 1999).

This is another intercellular pathway that, if interrupted, inhibits the ability of DCs to participate in the conversion of T cell to Treg. In this case, inhibiting p38 MAPK induction prevents DCs from converting naïve or anti-tumor T cells into Tregs. Inhibition of this conversion has its advantages but is still possibly problematic as it may increase the incidence of cancer-related anti-tumor mediated autoimmune reactions. On the other hand, control of this TLR activation pathway in DCs may be relevant to the manipulation of Tregs during immunotherapeutic treatment for cancer. If the treatment is unusually sensitive to Treg control, perhaps temporarily shutting down this pathway would allow for the treatment to achieve measurable effects, after which the pathway is allowed to function normally. In this way, inhibiting pathways that actually promote CD4 T cell effector development can make any Treg-sensitive immunotherapeutic treatment protocol more effective.

Blocking DC intercellular activation pathways and using anti-DC receptor blocking antibodies can interfere with Treg control of DC function

Since Tregs have the ability to control the activity of DCs, the therapeutic use of these cells must include inhibiting the ability of resident tumor-associated Tregs to prevent the DCs from performing their primary function, which is to induce an effective anti-tumor T cell response. Tregs chiefly control DC maturation and ability to present tumor antigens by making direct contact with the DCs. This occurs primarily in the draining lymph node nearest the tumor (Viguier et al. 2004). As discussed before, Tregs secrete a CD4-related transmembrane protein expressed from the lymphocyte activation gene-3 (LAG-3). The ligand on DCs is the MHC II molecule. This interaction results in the inhibition of DC activation, which is essential for the production of a productive anti-cancer T cell response. To prevent this interaction, it would be beneficial to mask the MHC class II molecule with a non–cross-linking antibody. But that would be impractical, because therapeutic DCs are supposed to present tumor antigens via MHC class II. Instead, these DCs can be pretreated with inhibitors that suppress ITAM, an inhibitory pathway. The pathway involves ERK recruitment of SHP-1 (Liang et al. 2008). Manassantin A has been shown to inhibit activation of NF-kappa B, ERK, and p38 (Kim et al. 2009).

The other aspect of Tregs controlling DCs through cell-to-cell contact has to do with surface proteins expressed by both cells. Tregs have been shown to associate with DCs on a preferential basis. They seem to be exceptionally mobile and find DCs immediately. This interaction occurs through the interaction with the LFA-1 molecule expressed by Tregs. Dendritic cells express the receptor ICAM-1. The interaction between these two represents the initial contact and results in the aggregation of Tregs and immature DCs. It also results in the down-modulation of CD80 (B7.1) and 86 (B7.2), which prevents DCs from activating tumor antigen–specific naïve CD4 T cells. This activity can result in tumor antigen–specific immune suppression and tolerance. Blocking LFA-1 with a non–cross-linking antibody could thwart the efforts of Tregs to induce inhibition of therapeutic DCs by down-modulation of CD80 and CD86 (Onishi et al. 2008). Of course, this treatment would have to be carried out in a systemic manner, but perhaps if it is done just prior to the infusion of therapeutic DCs, the system-wide effects of the antibody would be minimal. Another more sensible way to block this interaction would be to coat the immunotherapeutic DCs with a whole or F′ (ab) fragment of an anti-ICAM-1 antibody that blocks the LFA-1/ICAM-1 interaction. Since it takes some time for the DCs to recycle new ICAM-1 proteins, the cells would be resistant to this "instant" attraction by Tregs and naïve tumor-specific T cells could gather around the DCs in preparation for contact and subsequent activation by the immunotherapeutic tumor antigen–presenting DC. This means that infused DCs would be able to activate naïve anti-tumor T cells without interference caused by Tregs.

In conclusion, it is clearly important to control T regulatory activity during immunotherapeutic treatment of cancer. The treatments are based on how the T cell immune system arrives at its goal of attacking and removing a cancer. T cell memory generation is an essential component of this process. It is also apparent from the data derived from many clinical trials that Tregs not only interfere with the process, but also can render these procedures useless. Even though there is measurable *in vitro* and *in vivo* immune activity as a result of some of these clinical trials, there is no real regression of the tumor or long-lasting interruption of the steady progression of metastatic disease. This apparent dichotomy is a result of a sometimes overly efficient T cell immune system that controls autoimmune T cell responses via the Treg cell.

This activity during an anti-tumor T cell response not only inhibits the response but also allows the tumor to grow. This is a problem even without immunotherapeutic intervention, but with intervention, other issues may arise. Once immunotherapy is initiated, the activities of Tregs are possibly increased because of DC involvement in many of these therapies. Potentially the most unwanted side effect of using DCs as immunotherapeutic vehicles is their ability to convert antigen-specific anti-tumor T cells into tumor antigen–specific Tregs. This type of conversion, mediated by DC, has been shown to occur *in vitro* (Long et al. 2009) and *in vivo* (Cong et al. 2005). These immunotherapeutic DCs could potentially exacerbate the immune paralysis that may eventually lead to continued growth of the tumor, which is why the therapy had to be used in the first place. If therapeutic DCs actually begin to convert resident anti-tumor T cells into antigen-specific Tregs, not only does the therapy remove a much-needed source of anti-tumor T cells, but also any newly created or resident anti-tumor T cells would be unable to respond because of the increased numbers of Tregs brought about by the conversion process. To control the activities of Treg cells during immunotherapy, the activities of DCs related to Treg conversion must be understood and used so as to promote the anti-tumor T cell response initiated by the therapy. This type of conversion can be blocked using inhibitors of the intercellular pathways that lead to the conversion process. Systemic application of these inhibitors is not necessary because the DCs used for the therapy can be treated with inhibitors prior to infusion into the patient.

Dendritic cells efficiently exert their control over Tregs in many ways. Lymphokine secretion by DCs can be the result of maturation induced by the secretion of lymphokines by T cells or by cell-to-cell contact through a series of receptors and ligands expressed by both cells. The secretion of Treg-influencing lymphokines by DCs is quite complex, but it is a good target for the manipulation of Tregs during anti-tumor immunotherapy. During this activity, DCs release a host of powerful lymphokines, such as IL-6, IL-10, TGF-B, and IL27, that have complex immuno-modulating roles. The lymphokines control the differentiation of Tregs and other T cell subsets, such as Th-2 and Th-17 CD4 T cells. For instance, DCs can secrete IL-27 in response to beta-interferon (INF-b) and gamma-interferon (INF-g) by T cells. IL-27 was initially described as a lymphokine that induces differentiation (Szabo et al. 2000) and proliferation (Pflanz et al. 2002) of

Th-1 CD4 T cells. It was later found to play a role in the generation of Tregs by inducing STAT-1, which in turn binds to the FOX3 promoter and induces FOX3 expression and the Treg phenotype (Ouaked 2009). In support of the role of IL-27 in modulating T cell regulation, another study demonstrated that when DCs were exposed to INF-B they secreted IL-27 through a TLR-7 signaling pathway. This activity induced Tregs and also led to the reduction of IL-17, which can reduce the development of autoimmune Th-17 CD4 T cells, a situation that favors the development of Tregs (Zhang et al. 2009). However, in seemingly direct contrast to these studies, it was shown that IL-27 and IL-2 treatment of mouse neuroblastomas resulted in regression of the tumors and a reduction in tumor infiltrating Treg by inhibition of TGF-B secretion (Salcedo et al. 2009). The role of IL-2 in this IL-27-mediated process is essential. Despite this complexity involving DC-secreted lymphokines and the differentiation and/or conversion of Tregs, the fact is that these activities can be manipulated during anti-tumor immunotherapy.

Cancer immunotherapy is the next frontier in cancer treatment. Immunotherapeutic treatments are less toxic and painful than conventional treatments. Once perfected, they will increase the cure rate of many aggressive forms of cancer. A detailed understanding of how the T cell immune system functions when cancer is present is the key. Our T cell immune system is designed to recognize and remove a foreign threat. However, cancer is special because it originates from our own cells. There is a parallel T cell immune system (T cell regulatory), which safeguards against the development of anti-self (autoimmune) T cell responses that can be deadly. This T cell regulatory system composed of specialized T cells called Tregs holds autoimmune diseases such as diabetes, lupus, and arthritis at bay. The intersection between these two T cell systems (anti-cancer T cell and regulatory T cell) may dictate the outcome of most cases of cancer. Because cancer cells are derived from self, it is difficult for both these systems to act decisively at a critical time when cancer develops. Once this point is past without recognition and destruction of the cancer by the T cell immune system, it may be too late. The cancer continues to differentiate. It develops ways to thwart any attempt by the T cell immune system to destroy it. The interference of the anti-cancer T cell immune response by the T regulatory system allows the cancer to evade detection and outnumber the anti-cancer T cell pool. This is why it is critical to understand the dynamics of T cell regulation of the anti-cancer T cell immune response (Piersma et al. 2008). Over the next decade, we should find ways to separate the anti-self protective T cell regulatory responses that allow cancers to grow from T cell responses that attack cancer. Controlling the activity of T regulatory cells is essential to the development of successful anti-cancer treatments.

Chapter 12

NK Cells and Their Role in the Development of Immune-based Anti-cancer Therapy

Natural killer cells are an essential component of the immune system, but also an integral part of the anti-cancer T cell immune response. Humans who have a genetic deficiency in NK cells almost always suffer from fatal infections during childhood. The NK cell is a very different type of immune cell. It is thought that a cytotoxic effector cell, which would be a part of the primordial "innate" immune system, would be similar to an NK cell. This would have occurred prior to the development of T and B cells of today's "adaptive" immune system that has existed for 500 million years. Even though it has been proven that the NK cell of today developed after the existence of T and B cells, it still exhibits primordial functions. Initially it was thought that NK cells were cytotoxic only to cells that lacked self-MHC class I proteins. This was called the "missing self" theory of NK cell function. However, as knowledge grew about how NK cells recognized their targets, it was clear that they have many receptors that recognize MHC class I and MHC class I-like proteins, as well as proteins not related to MHC class I, which is why they have the ability to attack MHC class I negative targets. Because NK cells and T cells kill their targets by a similar mechanism (perforin), it is thought that these two cells developed from a common ancestor. Despite this similarity to T cells, NK cells express totally different cell surface proteins. For instance, they do not express the CD3 protein (Caligiuri 2008). As outlined in Chapter 9, this is an essential protein expressed by T cells because it is intimately associated with the assembly and function of the T cell receptor (TCR).

Natural killer cells provide an essential service for the anti-cancer T cell response, which is to provide immunogenic peptides that the T cell immune system needs to initially attack a forming tumor. These peptides are a result of macrophages and dendritic cells (DCs) processing the remnants of a tumor destroyed by the NK cell. This crucial function of NK cells is a function the anti-tumor T cell immune response cannot do without. Natural killer cells drift around the body and go places that traditional T cells cannot. They occasionally encounter a cell surface that appears different from normal. Technically these abnormal differences have to do with glycoproteins (carbohydrate and protein). These proteins on a normal cell surface are organized in a predictable pattern. When a cell starts to become malignant, they become disorganized and the NK cell can detect that change. Once the NK cells detect this difference, their receptors begin to aggregate, signals are sent to the cell nucleus, and there is an activation of the cytolytic cycle, which involves the induction of the perforin granuoles and the activation of the "death gene" in the target cell (Screpanti et al. 2005). In the end, the "suspect" cell is blown apart. The way cells die has a profound effect on the way the cellular antigens are processed by antigen-presenting cells (APC). Natural killer cells induce death through peptides (processed by APC) that are highly immunogenic to the T cell immune system (Moretta 2003).

Traditionally it was thought that NK cells were just effectors, but it was later found that a subset of these cells secreted lymphokines IL-4 and INF-g, which induce CD4 T cell activation and increase the expression of MHC class II proteins by tumor cells. This induces more tumor antigen–specific CD4 T cell activation (Strowig et al. 2008). This role, which was initially thought to be a T cell function, is dwarfed in comparison with the relationship that NK cells have with DCs. This is another situation where both cells control each other depending on the situation. In this chapter, we examine the role NK cells play in the immune system and why they are essential to the immune response to tumor cells. We address the intercellular events within the NK cells that impact the anti-cancer T cell immune response. Finally, we examine the role NK cells can play in the development of new anti-cancer immunotherapeutic treatment protocols

Role of NK Cells in the Everyday Working of the Immune System

Natural killer cells supposedly do not have a "classic" genetic memory like T cells, yet they are an important component of the immune system. They detect cell surface changes in other cells. When a cell becomes infected with a virus, the viral particles or proteins protrude through the infected cell's membrane. Natural killer cells can detect this protrusion and kill the cell before the virus completes its life-cycle. They are always on patrol, making contact with virtually every cell in the body. These cells live in the lymphatic system. Once they make their rounds, they park themselves in the nearest lymph node. Natural killer cells exist in two states, a resting state and an activation stage. One cell is responsible for making NK cells angry and that is the DC (Ferlazzo and Munz 2009).

When a virus invades or a tumor develops, NK cells attack the targets and the result is devastating. What is left behind is processed by macrophages and DCs. Once they kill their targets, NK cells can command other cells to stay behind and then leave the site. They move to their next target or go back home to a lymph node.

The NK cell is one of the most complex immune cells we have in our bodies. It's only now that a clearer picture of their nature has come to light. Several aspects of the NK cell immunology are controversial, starting with its very nature. The immune system is made up of two systems: the "innate" and the "adaptive". The innate system relies on genetic information stored in the genome. The TCR exists because it carries genetic information related to the recognition of bacteria, viruses, and tumor cells within the TCR genes. This information can be called upon at a moment's notice to marshal the forces of the immune system to attack those threats. Now, the adaptive immune system relies on the mutation of antigen-specific receptor (T cell or B cell) genes to conform to a threat of which there is no genetic memory. The NK cells were thought to be a part of the innate system because they supposedly have no memory of their encounter with their target but are predisposed to recognize it. In 2009 this concept was challenged when scientists found that when NK cells responded to a virus they expanded 100-fold. Once the viral threat disappeared, the expanded NK cells went through a "contraction phase". During this phase, they could be re-activated to respond if the virus reappeared (Sun JC et al. 2009). The author of that study considered this activity "adaptive", but is it really?

Another controversial aspect of NK cells is their similarity to T cells. It has been determined that some NK cells do express a TCR chain. This chain is the alpha chain but, unlike the one expressed by T cells, it has no partner. In T cells there is an alpha and a beta chain, which combine on the T cell surface forming a binding groove where the immunogenic peptide of the target cell gets positioned. The TCR alpha chain expressed by this subset of NK cells has an entirely different function. The NK cells that express this TCR alpha chain are thought to be special. They protect patients who receive bone marrow transplants from the devastating effects of graft vs. host disease (GVHD). This is one of the major phenotypic characteristics NK cells share with T cells. When these NK cells were discovered they were thought to be a new type of T cell. Careful study of the cell surface revealed that they were indeed NK cells but expressed a TCR alpha chain (Kronenberg 2005).

Natural killer cells reside in the lymph nodes, where they were initially thought to be resting, but subsequently found to be quite mobile, with the same amoeba-like movement characteristic of T cells and B cells. They move from place to place making contact with the other immune cells and particularly with DCs. Natural killer cells express a protein on their cell surface called CD49b. Cross-linking this protein on the cell surface halts all NK cell motility and, it is believed, mimics what happens when NK cells reach collagen-rich sections of the lymph node. In these areas of the lymph nodes there are APCs (DCs) communicating with T cells. It is theorized that NK cells that stop in these areas begin to secrete INF-g. That secretion induces MHC class II expression by APC, thereby allowing these cells to present even more antigens to antigen-specific T cells (Garrod 2007). In the same study, it was discovered that NK cells also have a relationship with B cells. They can swarm to the location of target cells, destroy them and make contact with B cells. Once this contact is made, the B cell cannot move. The idea here is that the NK cell makes sure that B cells stay near the dead target cells so they can present tumor or viral antigens to T cells. In response, the T cells secrete B cell-specific lymphokines that induce the B cells to make tumor or viral-specific antibodies (Garrod 2007). Once these antibodies attach to the target cell, the NK cell can be called back to kill more virus-infected cells or tumor cells because of its Fc receptor affinity for antibody-coated cells. This is an important role the NK cells play during the development of a tumor and during an initial viral infection where infectious viruses escape from the lysed target cells.

These are many ways in which NK cells support the immune system during routine infections and the development of malignant cells. They seem to be able to bridge the gap between the innate and adaptive immune systems. They are among the "first responders" of the cellular immune system. When a new cancer develops, they act immediately and swiftly. They can also cause other immune cells to follow them or remain behind to begin the process of a full-blown immune response involving just about every type of immune cell in the body. The NK cell is one of the most complex immune cells. This complexity allows it to function in ways that make it an irreplaceable component of the immune system (Terabe and Berzofsky 2008).

Why NK Cells Are Essential to the T cell Immune Response to Cancer

The first and possibly most important contribution NK cells make to the anti-cancer T cell response is that they provide APCs with much-needed immunogenic tumor proteins. The proteins are processed into immunogenic peptides and presented to tumor-specific CD4 and CD8 naïve T cells. Natural killer cells kill tumors by either secreting perforin or by inducing a "death gene". In the perforin system, perforin granules inside NK cells gather at the synapse (point of cell-to-cell contact between the NK cell and its target). The perforin is released through the NK cell membrane, makes contact with the target cell membrane, and then enzymatically bores a hole in the target cell membrane. After a number of these pores are formed, the cytoplasm of the target cell spills out into the environment and the cell dies. In the second method, NK cells induce a "death gene" within the target cell. There is a receptor/ligand system called the Fas system (Screpanti et al. 2005). This is a somewhat primordial method the immune system has used to control the growth of not only cancerous cells, but also normal cells that continue to grow, for example, in the regeneration of liver cells (Nagata 1996). The Fas system has two parts. The first is a receptor (FasR) that is expressed by many cells in the body and can be induced in many more by various means. Once a Fas ligand (FasL) makes contact with the receptor, a series of events occur that cause the DNA of the cell to fragment and lead to immediate cell death. Natural killer cells express the Fas ligand. Once this protein on the NK surface makes contact with the FasR on the target cell, the death of that cell is all but certain (Screpanti et al. 2005). The cellular peptides that result from this type of target cell death are not as immunogenic as those produced by the perforin system. This type of tumor cell death is called apoptosis and for some reason the antigens made available to the T cell immune system (via the APC) are from the inside of the tumor cells and often induce autoimmune T cell responses instead of anti-tumor responses directed at the tumor cell surface (Navratil et al. 2004). The end result, though, is that there is one less living tumor cell in the body.

Another important and recently understood aspect of the NK/T cell relationship is related to the generation of an anti-cancer T cell immune response. Subsets of NK cells release lymphokines. A specific function that applies to anti-cancer T cell responses requires the help of the DC. During an active T cell immune response to a growing tumor mass, DCs are created by the secretion of IL-4 and GMCF by tumor-responding T cells. They sample the tumor antigens from the dead cells in the tumor mass and make their way to the lymph nodes, where they make contact with NK cells. The NK cells make their way to the tumor mass and kill tumor cells. Before the NK cells leave the lymph node, the DCs instruct a subset of these cells to secrete IL-2 (Terabe and Berzofsky 2008). IL-2 is used by precursor CD8 T cells to develop into cytotoxic T cells capable of killing the tumor. So, not only do NK cells attack the tumor, but they also activate the tumor-killing function of CD8 T cells.

Among the lymphokines that NK cells release is the INF-g, which has a profound positive effect on the anti-cancer T cell immune response. A subset of NK cells called CD56++ (versus just CD56 positive) secretes INF-g. They need two signals to do this. The first signal is the lymphokine IL-12, and the second signal comes from either lymphokines (IL-1, IL-2, IL-15, or IL-18) or the engagement of the NK cell's Fc receptor by antibody-coated cells. The lymphokines can be secreted by monocytes, macrophages, or DCs. This is another way the cells of the monocyte/ macrophage series control NK function. They release INF-g more quickly than T cells because it takes several rounds of division for T cells to release INF-g. The lymphokine contributes to the acceleration of the anti-tumor T cell response (Strowig et al. 2008) because it up-regulates the expression of both MHC class I and class II proteins on APCs. This accelerates the anti-tumor T cell response because it means that APCs are able to present even more immunogenic peptides to both CD4 and CD8 T cells. The secretion of INF-g by NK cells also has a built-in safeguard to diminish the response after the tumor cells are killed. Continued exposure to INF-g causes APC to secrete IL-10 and TGF-B. These two lymphokines turn off INF-g secretion by NK cells, inhibit T cell proliferation, but have also been shown to induce the formation of Tregs (Jonuleit et al. 2000). The Tregs, through their ability to stop T cell responses, effectively end any activity associated with the T cell response to a tumor threat that no longer exists. This is essential to prevent cancer-induced autoimmunity (as you will remember from Chapter 7).

Intercellular Events within the NK Cell that Have an Impact on Anti-cancer T cell Response

Because of the complex nature of the NK cell it is one of the last immune cells to have its intercellular pathways understood. The NK cell itself is quite basic in what it does, but the biochemical reactions inside it are complex. Those reactions, which include the intercellular pathways, intersect with a higher frequency than in T cells. This was a somewhat unexpected finding, from the perspective of immunobiology, because the T cell not only differentiates into many subtypes, but also secretes a host of lymphokines, cytokines, and chemokines. Because of the complexity and multiple intersections of the NK cell pathway system, scientists have understood only a small fraction of the pathways. Among the intercellular pathways that are understood are some that are unique to the NK cell. Some of them are related to some common intercytoplasmic proteins, such as kinase substrates (ITAM) and MAP kinases (syk and phospholipase C). Because of this, they are sensitive to well-known inhibitors utilizing the intercytoplasmic proteins. The last section of this chapter will discuss possible alterations in these pathways that should make NK cells more effective in anti-cancer immunotherapy.

The next intercellular pathway involves the ability of NK cells to become activated and kill their target. When activated DCs make contact with NK cells in the lymph nodes after leaving a tumor site, the NK cells are induced to secrete IL-2. Once the NK cells leave the lymph nodes and travel to the tumor site, they activate precursor CD8 T cells to mature into anti-tumor cytotoxic CD8 T cells. However, IL-2 affects not only the immune cells (CD4 and CD8) around the NK cell, but also the NK cell itself. This is called "autocrine" stimulation: the substance the cell secretes activates the secreting cell. Part of this activation of NK cells by IL-2 has to do with an increased ability to kill its target. This is accomplished by the ability of IL-2 to mobilize phosphatidylcholine-specific phospholipase C, or PC-PLC, which usually is dispersed throughout the cytoplasm of NK cells. IL-2 activation causes PC-PLC to move toward the cell membrane. During this phase, there is a calcium influx and some kinases from the syk family such as Zap-70 are activated. The result is a concentration of granules containing perforin, which is used in the target-killing phase. After IL-2 activation and PC-PLC enzyme activity, the granules begin to travel down microtubules toward the cell membrane and position themselves for fusion with the membrane and subsequent exposure to the target cell membrane. The PC-PLC enzyme makes this possible because inhibitors of the intercytoplasmic protein completely interrupt not only anti-tumor killing, but also NK cell activation. The regulation of perforin granules by PC-PLC is an essential intercellular function. Without it, the T cell immune system may not be alerted to the presence of a developing tumor until it is too late (Ramoni et al. 2001).

The next example demonstrates how complex intercellular pathways intersect during the function of NK cells to reinforce a desired activity. Earlier we discussed how a surface protein on NK cells controls its transit through the lymph nodes. This was the CD49b protein, and it is called an integrin. It is made up of two chains, an alpha and a beta chain. Integrins are expressed by many cells and are involved in cell adhesion and migration of cells from one point to another. Once the proteins are cross-linked by the collagen-rich areas of the lymph node, NK cell motility is arrested and the cells begin to secrete INF-g, but the cells also are activated to kill targets through the intercellular pathway mentioned above. However, the cross-linking of CD49b induces the activation of H-Ras (Garrod 2007), which is an essential part of T and NK cell function. In a chain reaction, H-Ras activates MAP kinases, which activate ERK, which in turn activates (phosphorylates) AP1 (Marusina et al. 2008). AP1 is located within the nucleus and signals it to increase the expression of the IL-2 receptor on the cell surface so that IL-2 can induce the PC-PLC enzyme to mobilize perforin granules to the NK cell membrane, which prepares them to kill a target such as a newly formed tumor cell. This is a perfect example of how the intercellular pathways in NK cells intersect and amplify each other. The biochemical reactions that take place during these events happen within seconds of the stimulus that caused them. The NK cell is as much a biological machine as the T cell, but absolutely none of these intercellular activities require the NK cell to divide. This is all programmed into the genetic make-up of the cell. However, its receptors command a powerful series of intercellular events that involve a direct line of communication with its nucleus.

The nucleus of an NK cell is constantly signaled by its surface receptors. These receptors are unusual for a T cell. Several receptors work in reverse of the "classical" TCR model: inhibitory killer immunoglobulin-like receptors or KIRs, and inhibitory CD94-NKG2A heterodimeric receptors. There are variants (slightly different protein sequences) of each one of these receptors. Each NK cell expresses a combination of variant receptors. For a T cell immunologist, this sounds like Aramaic. Natural killer cells are rudimentary in their recognition of target cells. When an NK cell surveys a normal cell that expresses a normal MHC class I protein, it "latches on". The KIRs induce signals that override any NK cell activation signal produced. Therefore, the NK cell detaches and moves on. The same thing happens with the CD94-NKG2A receptor, but it is associated with MHC class Ib, which is called HLA-E (Caligiuri 2008). For the NK cell to kill a target, it must see one of these MHC class I proteins and a "messed-up" glycoprotein version of either. The "messed-up" versions are a result of either viral infection or altered cell surface proteins that appear on the cell surface of a newly developed tumor. Once these requirements are met, the NK cell decimates the target cell. However, several other NK cell receptors (NKG2D and NCR) see non-MHC class I proteins and can, upon engagement with these proteins, override the inhibitory signals of the KIRs and CD94-NKG2A and induce NK cell activation and killing. These antigens seen by these two receptors (especially NKG2D) have some MHC class I homology but are not expressed on normal tissues. However, they are expressed on cells undergoing DNA mutations (genotoxic) or cellular stress (Raulet and Guerra 2009). Therefore, the NK cell, because of its receptor configuration, can kill a tumor or virally infected cell that is barely (or not) expressing MHC-class I proteins.

These NK cell receptors initiate signals through ITAM-containing adapters such as the Fc domain and DAP12 (DAP10 in humans). These adapters send signals directly to the NF-kB complex. After proteolysis of NFkB, a positive signal is sent to the nucleus (Marusina et al. 2008). In order for the above adapters to activate NFkB, they must make contact with a protein called caspase recruitment domain 9, or CARD9. This pathway is important because it controls not only the ability of NK cell to kill targets, but also lymphokine and cytokine secretion by these cells. It is a unique pathway used this way only by NK cells. This pathway is used by more than five receptor/ligand systems in NK cells, whereas T and B cells use it with only two systems. In this way NK cells are distinct from all other immune cells that participate in the anti-tumor T cell response.

This finding has some interesting implications because protein kinase C is involved (Hara et al. 2008). Inhibitors of such kinases affect only the ability of NK cells to secrete lymphokines. That is of great interest because of the role NK cells can play in the inhibition of GVHD during bone marrow transplant therapy. This disease can be deadly to cancer patients. Natural killer cells can minimize or potentially exacerbate GVHD depending on the type of lymphokines released and the time of their release during therapy. Manipulation of this activity by selective inhibition of these kinases could potentially prevent GVHD from occurring. This will be covered in detail in the next section on the use of NK cells during anti-cancer immunotherapy.

The Role NK Cells Can Play in the Development of Anti-cancer Immunotherapeutic Treatments

Natural killer cells play a special role in immune-based treatment of cancer because of their role in cellular immunity in general and anti-tumor T cell immunity in particular. Several aspects related to their function make them ideal candidates for use during therapy: their immune surveillance function and their ability to attack and kill MHC-negative targets. Because NK cells have this special ability to kill tumor cells, they are perfect for eliminating leukemic cells because these leukemic cells down-regulate the expression of both MHC class I and class II. The other fortunate thing about leukemic cell antigen expression is that leukemic cells almost always express the NKG2D receptor antigens that are expressed during genotoxic or cellular stress. This gives the NK cell an advantage over the T cells in these situations because T cells need expression of MHC proteins to respond to the tumor. The other major advantage NK cells have over T cells in a transplant situation is that they are designed not to attack normal tissues by virtue of their receptor function. These last two differences between T cells and NK cells allow the NK cell to be used to prevent GVHD and also to participate in the graft vs. tumor (GVT) response (Sutlu and Alici 2009).

Let's step back and try to define transplantation treatment for cancer. There are many sources of transplanted cells and complications involved with each procedure. There are several types of transplant procedures, depending on the type of cancer. One procedure is the bone marrow (BM) transplant. Variations of this procedure have to do with the source of cells and the pretreatment of the donor cells prior to infusion into the patient. The cells of the donor and the recipient must be typed for MHC proteins prior to transplantation. This is what is called a haploidentical transplant, where the BM comes from a sibling or close relative, and the MHC proteins match in an almost identical fashion. The operative word here is "almost".

The idea is that the T cells in the donor's BM are so closely matched to the recipient's tissues that they see them as self. The leukemia, however, is something the donor's T cells have never seen and they proceed to kill it. Sometimes, after the leukemic cells are destroyed, the donor's cells take up residence in the recipient's body. This is called chimerism. The recipient has to be conditioned prior to the transplant procedure. This usually entails radiation and chemotherapy to remove all the lymphocytes and leukemic cells circulating in the patient's body. This is called lymphoablation, and it leaves the patient without an immune system until the BM produces more cells, which unfortunately means leukemia cells as well (Stackstein 1995).

Graft vs. host disease becomes an issue because even if the donor and recipient are an identical match (such as twins) there are still minor differences in their MHC proteins that the donor's T cells can recognize. These are called minor histocompatibility antigens (MHA). The MHA are derived from subtle changes in DNA after we are born. The DNA changes sometimes result in a change of one or two amino acids in a tissue-specific peptide, but they are enough for the donor's T cell to recognize as foreign and attack that tissue (Kernan and DuPont 1996). The attacked tissue may be a major organ such as the liver, intestines, or skin. These GVHD reactions can be deadly in cancer patients weakened by the radiation and chemotherapy that were necessary for the transplant procedure. Once the reactions start they are very difficult to stop. It is necessary to use powerful anti-rejection drugs to diminish the effects of GVHD, but the goal is to find a way to avoid it. The disease starts with the recipient's DC. Most cells in our bodies carry the same genetic information, and therefore the same type of proteins. So the DC would carry the same MHA that a liver cell or skin cell would. However, the DC is the most powerful APC in our body and when the haploidentical T cell interacts with it as it does with its own DC, but sees the MHA, the T cell becomes activated, CD8 T cells become anti-MHA killer cells, and the destruction of the recipient's organs begins. Therefore, the recipient's DCs are the focal point for the development of GVHD (Ruggeri et al. 2002).

There is a variation of the BM transplant technique for treating leukemia. It involves the removal of T cells from the donor's BM prior to transfer into the leukemic recipient. This process does not remove every single T cell. In addition, the transplanted BM cells produce precursor T cells of the donor. T cells are born in the BM and develop their tolerance of self and recognition of non-self in the thymus. The idea here is that the BM of the donor takes up residence in the recipient; pre-mature T cells are formed and make their way to the recipient's thymus, where they mature. The mature T cells leave the thymus and do what the non-T cell depleted BM cells are supposed to do—attack the leukemia cells of the recipient. This time, though, the T cells that have matured in the recipient's thymus do not see the MHA because they have become tolerant to them. It was found that this procedure not only increased the GVT effect, but also resulted in a much lower incidence of GVHD. For some time the logical assumption was that the lack of GVHD was indeed caused by this elimination of T cells in the thymus that recognized MHA. However, experimental evidence in mice revealed that the lack of GVHD and possibly the increased induction of GVT activity were due to not the thymus-selected T cells, but the donor's NK cells. This revelation started a flurry of scientific research that confirmed that NK cells of the donor were largely responsible for both activities associated with successful T cell-depleted BM cell transplant therapy for leukemia.

The evidence for this was sound, and cellular immunologists clearly demonstrated that the NK cells were indeed responsible for the success of the use of T cell-depleted BM cells from a haploidentical donor. They contributed to the successful treatment of leukemia and were responsible for the absence of GVHD complications (Haraguchi et al. 2005). The experiments were designed to demonstrate how donor T cells, once instructed by the recipient's thymus not to react to the MHA of the recipient's tissue, would attack the leukemic cells and ignore the normal tissues. However, it was the donor's NK cell that did both things and did them remarkably well. First, NK cells kill leukemia cells with extraordinary impunity (Velardi et al. 2009). In some respects. the leukemia cell is the perfect target for NK cells. The primary reason leukemia results from BM cells that would become either T cells or myeloid cells (mast cells, polymorphonuclear cells) is that they develop massive DNA mutations that result in cell surface protein disfiguration. Some of these mutations are caused by external stimuli, but some are caused by inherited genetic abnormalities like the Philadelphia chromosome discovered by Dr. Peter Nowell of the University of Pennsylvania (Nowell et al. 1984). This is a type of cellular stress that can induce the NKG2D antigen expression on leukemia cells and is definitely due to a genotoxic event. This process of DNA mutation that creates leukemic cells not only severely changes the MHC class I proteins, but inadvertently produces the non-MHC antigens recognized by the NKG2D and other NK receptors (Raulet and Guerra 2009). This signal activates the NK cell along with its ability to kill the cell.

Why do the donor's NK cells prevent GVHD? This function is actually quite simple because of the NK cell's ability to quickly recognize and react to altered MHC class I proteins. In this situation the MHC class I proteins are different from those of the donor. When totally haploidentical donor BM cells are used during the transplant, the MHA are presented to T cells by the recipient's DCs. This occurs during traditional BM therapy for leukemia. It induces a series of T cell-related events that eventually lead to GVHD. This takes time because there is presentation of MHA, several rounds of T cell division, and induction of CD8 T cell precursors, and then the attack on normal tissues commences. The cell that is at the focus of this induction of GVHD is the DC. The DC of the recipient expresses MHA that represent the entire organ system of the recipient and these are presented to the donor's T cells. The T cells see these peptides and begin their methodical response to find and damage the tissues that express them. The odd thing is that the donor's NK cells see the same peptides associated with the MHC class I proteins of the recipient's DC. They see them not as antigen, as T cells see them, but as altered MHC class I proteins. The NK cell sees the MHA differences between the donor and the leukemia patient recipient and also tissue-specific peptides presented by DCs of the recipient. The donor's NK cells kill as many of the recipient's DCs as possible. Once the NK cells evacuate the recipient DCs, the ability of donor T cells to get immunized against recipient tissues diminishes. Therefore, the donor's NK cells can attenuate the development of GVHD not only during BM T cell-depleted stage, but also during traditional BM therapy for leukemia (Ruggeri et al. 2002).

There are two more variations of the transplant procedure for the treatment of leukemia and possibly other forms of cancer. Both procedures make use of NK cells to again attack the tumor and reduce GVHD. The first uses the same haploidentical T cell-depleted BM, but in addition, alloreactive NK cells are added to the donor BM. The alloreactive NK cells (from an unrelated donor) cannot recognize the MHC proteins of the recipient's normal tissues because NK cells are programmed not to attack normal tissue. However, they can react to DCs presenting antigens, which are recognized as altered MHC antigens. The alloreactive NK cells promptly kill those DCs and make them unavailable to GVHD-inducing T cells. This is how they prevent GVHD, but they do it more efficiently because of the alloreactive nature of the added NK cells. In cases of leukemia and other tumors, the NK cells see the same non-MHC antigens, but again they are much more effective at eliminating the tumors than NK cells derived from the donor (Lundqvist et al. 2007). This use of NK cells during BM transplants is a promising method to augment cancer immunotherapy. The second variation of the transplant process is the use of stem cells. Stem cells, like BM cells, should be matched to the recipient, but they really don't have to be because they can take on the characteristics of the recipient's MHC protein configurations. Again, alloreactive NK cells have been added to the transplanted cells with the same result—a reduction in the tumor and of GVHD (Bertaina et al. 2009).

The issue of NK cells attenuating GVHD during BM transplant therapy for leukemia is complex. For what seems to be a simple immune cell, the NK cell plays a complex role in reducing GVHD. Some subsets of

NK cells share characteristics with T cells. The TCR alpha chain can cause problems during the depletion of T cells from the donor's BM. When T cell-depleted bone marrow cells are prepared for alloreactive transplant, the TCR alpha chain NK cells (NKT cells) get removed as well. These cells can play a major role in reducing GVHD. They do this by two methods, which simultaneously attenuate GVHD. First, the donor NKT cells become activated. This leads to an increased expression of the Fas ligand. The donor NK cells begin to kill donor CD4 T cells by FasL/FasR induction. This activity reduces the number of CD4 T cells that potentially could be induced to damage tissues of the recipient. The process initially slows down, but it's what the NKT cells do next that may ultimately prevent GVHD in the first place. In the second method, the activated NKT cells secrete the lymphokines IL-4 and INF-g in large quantities (Kim et al. 2007). We have discussed the immune-modulating effects of INF-g (increasing MHC expression in APC and tumor cells) and IL-4 (used with GMCF to induce DCs), but we have not discussed what they do when secreted in combination. The two major types of T cell responses are Th-1 (induction of cytotoxic CD8 T cells) and Th-2 (induction of antibodies secreted by B cells). When IFN-g and IL-4 are secreted together, they create conditions in which a Th-2 T cell response occurs. The GVHD response by T cells is a Th-1 response because it involves the recruitment of CD8 cytotoxic T cells to destroy the normal tissues of the recipient. In this situation the donor's NK cells play a major role in preventing GVHD because they skew the donor's T cell response toward one that has little effect on the normal tissues of the recipient (Haraguchi et al. 2005).

In conclusion, NK cells are thought to be part of the innate cellular immune system because they are programmed to respond to threats such as viruses and tumors. This is largely true. However, NK cells have the ability to specifically interact with other cells of the immune system and modify their behavior. They exist in our bodies to automatically respond to perceived threats and to support both the innate and adaptive arms of the T cell immune system. In some respects NK cells have qualities of both innate responsiveness and adaptive behavior. This combination of functions allows the NK cell to bridge the gap between the two systems. The innate T cell response sees an immunological threat through the vast library of genetic information stored in the immunity genome library, which for all intents and purposes are the genes of the TCR. If, by chance, the threat does not express peptides found in this library, it is the job of the adaptive T cell immune response to deal with the problem. The adaptive T cell system can, through vigorous rounds of cell division, model a TCR through somatic mutation that will conform to the peptides of the threat, but it takes time and a non-specific T cell stimulus to induce this process. This is where the NK cell comes in. It can immediately begin attacking the threat on the basis of criteria that the T cells are not prepared to deal with. Natural killer cells provide much-needed peptides via the APCs and affect dormant T cells by the elaboration of lymphokines and chemokines. These can serve as non-specific stimulators of T cell proliferation, which will begin the somatic mutation process of the TCR and eventually produce a TCR that can see this previously unseen threat. Ultimately, the NK cells give the T cell immune system the extra time to coordinate and focus its efforts on destroying a newly formed cancer.

One of the chemokines released by the NK cell during its attack on a newly formed cancer is called CCL5. This chemokine is released simultaneously with the release of perforin from the cytotoxic granules during NK cell target killing. As we have said, NK cells usually do not hang around after they have decimated a target cell, but they do things to other immune cells to attract them or keep them at the site they just left. CCL5 attracts other immune cells like T cells, macrophages, and DCs to the site where the NK cells killed their targets and left. The discovery of this NK cell function has important implications for the anti-tumor T cell response (Kumar et al. 2009). First, NK cells kill tumor cells. Then they send out a soluble signal to the very cells that initiate an anti-tumor T cell response. Second, this activity, controlled by intercellular activation events, may be the key to the NK cell's "immune-surveillance" function. This is another mechanism ripe for manipulation during the immunotherapeutic treatment for cancer. If it is possible to amplify this chemo-attractant signal it may be possible to recruit larger numbers of T cells, macrophages, and DCs to an emerging tumor site. That means destroying cancer cells far in advance of their ability to induce protective T regulatory cells (Tregs) that would inhibit the generation of a productive anti-tumor T cell response. This concept applies to the early formation of cancer cells. The longer the tumor survives these initial stages, the greater the likelihood that it will defeat the T cell immune system and lead to death.

We earlier described granules containing perforin. This had to do with the ability of the PC-PLC enzyme to mobilize these granules to the inner-cytoplasmic region of the NK cell during activation. It turns out that the CCL5 vesicles follow the same path, which means that they too are mobilized by the PC-PLC enzyme. Unlike T cells, which have to develop into CD8 killer T cells and then release CCL5 after killing of a target, the NK cell already has CCL5 stored in its cytoplasm and continually makes the protein, evidenced by the constant presence of CCL5 mRNA. Again, unlike T cells, the CCL5 is released immediately after killing its target. Therefore, there has to be an intercellular activation pathway that will release the mobilized vesicles, not into the target cell, but into intercellular space. This pathway is among the familiar MAP kinase (ERK, JNK, and p38) pathways that also control the PC-PLC mobilization of perforin granules. However, only a specific pathway (JNK) of this series controls CCL5 mRNA levels as well as its promoter (SP1/KLF13), which controls the steady-state levels of the CCL5 protein. JNK stabilizes the interaction between the promoter and the CCL5 gene. Inhibition of protein kinase C (MAPK) reduces the ability of NK cells to release perforin (cytotoxicity) and also CCL5 (chemokine). Inhibition of JNK has no effect, because JNK only stabilizes the production of CCL5 (Kumar et al. 2009). If you increase JNK function (through protein transfection) then there should be an increase in CCL5 stability and quantity during NK cell killing. If more CCL5 is released during these periods, immunotherapeutic donor or allospecific NK cells added to BM transplants can alert an increased number of donor T cells, macrophages, or DCs to respond to what is left over from the NK cells' attack on leukemic cells.

Natural killer cells can be genetically modified for use in immunotherapy through the introduction of a receptor system designed to attack lymphoma cells that express the protein erbB2. The authors of a mouse study built a "chimeric" receptor that recognized the erbB2 protein. This receptor was shown to be stable and functional in its ability to overcome the inhibition of the KIRs and NKG2D receptor systems. These systems function to diminish the ability of NK cells to kill self. Once the transfected NK cells interacted with their target, they were activated and they secreted lymphokines typical of a responding NK cell (Pegram et al. 2008). This is an important study because the basic function of NK cells is quite simple and a procedure that can manipulate NK cell function has to be considered seriously.

Another approach related to targeting NK cells relies on the type of tumor and the NK cell-recognizable antigens they express. It was found that human melanoma cells were recognized by the expression of three NK cell receptor-recognized proteins. Two of these proteins, expressed by the melanoma cells, resulted in the death of a vast number of tumor cells when matched with NK cells that had that specific receptor. However, those NK cells were from other donors (mismatched). The NK cells from the patient did not kill the melanoma because of KIRs (Igarashi et al. 2004). These receptors override activation signals derived from the recognition of self non-MHC antigens. The use of NK cells from another donor negates this effect and the allospecific NK cells destroy the melanoma targets. In this way NK cells that have receptors for a particular melanoma cell can be used therapeutically.

These are just a few of the many recent developments concerning the role of NK cells in the development of immune-based anti-cancer therapy. Since NK cells have a limited ability to differentiate, it is quite easy to alter their function in a stable way. These cells are highly programmed. Alterations tend to be more stable than those performed on T cells.

T cells have to divide and differentiate in order to do what they do. Similar genetic alterations of T cells can be temporary because of the changing nature of the T cells during an immune response. In other words, the genetic alterations (insertion of a new gene) get lost or excised from the genome during the rapid cell division typical of T cells in response to an immunological threat such as cancer. The use of NK cells during immunotherapy is a logical choice because they are programmed and thus in a way single-minded. However, because of the intercellular complexity involved in NK cell function it may take some time to fully exploit the potential of these very important immune cells. The good thing about this situation is that NK cells can be cloned and expanded in the laboratory. Because of their pre-set nature, they undergo very few changes during replication. This is the case with T cells during *in vitro* cell proliferation. In this case, they often lose the natural gene expression of adhesion proteins after *in vitro* cultivation. The loss of these proteins means the immune T cells cannot adhere to the proper tissues needed to get them to the tumor mass. Natural killer cells tend to not lose these adhesion proteins because they do not differentiate *per se*, but basically copy what a particular NK cell is expressing at the time.

In this conclusion we have introduced how immunobiologists used NK cells and their non-conventional way of attacking tumor cells, especially the way they tried to tailor that type of NK cell recognition to either particular tumors or the antigens those tumor cells expressed. This is new information that may use the enormous potential of NK cells to not only destroy cancer cells, but also alert and remodel the T cell immune system to remove the cancer for good. This ability of NK cells to skew or tilt the balance of the T cell immune system is just being recognized and may turn out to be a fundamental breakthrough for anti-cancer immunotherapy. Once thought of as rudimentary and good for only immune surveillance, NK cells are now, with all the high technology we have at our command, viewed as a mystery, containing unknown functions and roles related to the immune response to cancer. Because of the ability of immunologists to marshal this new technology, it may be possible to use NK cell immunobiology to benefit the development of integrated approaches to the immunotherapeutic treatment of cancer. Then, and only then, will the NK cell be able to take on the role of the champion in anti-cancer immunotherapy.

Chapter 13

How Cancer Cells Influence the Anti-cancer T cell Immune Response and Why This Is a Problem during Immunotherapeutic Treatment of Cancer

The cancer cell (tumor cell and cancer cell will be used interchangeably throughout) itself is a vital element in the T cell response against it. The relationship between the cancer cell and the T cell is complex (see Chapter 3) as far as an active immune response is concerned. The complexity extends to the response and the systemic conditions created by the cancer cell during its development and growth. Those systemic conditions affect other systems directly. However, indirectly some of the changes created by the cancer make it much more difficult for T cells to directly attack the cancer. The cancer cell can create diversions, which misdirect the T cell response in ways that are harmful to a patient. The cancer creates conditions that weaken the entire biological system.

To treat cancer, especially with immunotherapy, these issues must be addressed. Systemic problems created by the cancer that indirectly affect the anti-cancer T cell immune response must be successfully treated as well. Without such an approach, new immune-based anti-cancer therapies are doomed to failure. Treatments successful in the lab and on mice have failed in human clinical trials.

In this chapter we discuss how the cancer cell is formed and how it can influence the anti-cancer T cell immune response. To a certain extent, the cancer cell can trick the T cell immune system into responding, only to damage it to the point where it becomes useless. Events that occur at the early stages of tumor development may involve less than a hundred tumor cells that set the stage for the T cell response. We will discuss the mechanisms involved, which are an important phase of the anti-cancer T cell response.

A developing cancer on its own can facilitate the type of T cell response that can completely destroy the cancer so that it never recurs. Some cancer cells can draw in the T cell immune system. The cancer can convert some of the T cells at the cancer site into cells that will protect it from attack. It can also weaken other biological systems in the body. This is an important concept to understand because some of these actions can lead to the death of the patient from secondary infections. These infections are caused by a form of immunosuppression induced by the cancer. The other potentially lethal aspect of cancer controlling the T cell immune system is its ability to redirect that system so it attacks self. This is cancer-induced autoimmunity (see Chapter 7) and can be just as deadly as the cancer itself. In the end, in order to successfully cure cancer, these issues must be resolved by treating the root causes of cancer-induced changes in our T cell immune and biological systems.

How do cancer cells influence the anti-cancer T cell response? We will outline the development of a tumor from a normal cell as it relates to its interaction with T cells as well as the immune cells that support them such as macrophages, dendritic cells (DCs), and natural killer (NK) cells. Within this section, we highlight activities related to the induction of T cell suppression, misdirection, and inhibition of T regulatory cells or Tregs. These cells (except for NK cells) are highlighted because of their role in the induction of a false positive anti-cancer T cell response and their possible contribution to the subsequent failure of that response. We see how an established cancer alters the T cell response in ways that make it ineffectual against cancer. We look at what happens when the cancer secretes T cell-recognizable peptide antigens, and the ability of cancer cells to secrete immune-modulating lymphokines that alter T cell and immune cell function. We also discuss how cancer cells weaken the entire biological system and affect the immune response. The results of the systemic effects brought about by the cancer, such as infections, will be covered.

Cancer-induced autoimmunity will be covered briefly since it was explained in Chapter 7. However, the aspects of autoimmunity that lead to a paralysis of the T cell immune system will be highlighted. One of the causes of cancer-induced anti-cancer T cell immune paralysis is the generation by the tumor of cancer-specific Tregs. We discuss this in detail because it is a powerful tool the cancer uses to avoid being killed by anti-cancer CD8 killer cells. We outline how cancer cells influence, inhibit, or interfere with newly developed anti-cancer immunotherapeutic treatment procedures. Finally, we discuss ways in which scientists are currently addressing these issues.

How Development of a Cancer Cell Influences the Anti-cancer T cell Immune Response

We have earlier explained in detail how a cancer develops from a normal cell, but specific aspects of this development are critical to the survival of any person who develops cancer. Many times in our life, we develop potentially cancerous cells, but our T cell immune system removes them before they can even be detected. The cancer cells get killed and an immunological memory is left behind in the form of memory T cells. Therefore, it is only when this amazing protection by our T cell immune system fails that we even discover the presence of the cancerous cells. That does not mean that the T cell immune system all by itself can at a later date remove the cancer. This process is called spontaneous remission (Rieger et al. 2009). It means that the longer the tumor remains intact, the higher the likelihood that it will cause serious problems for the patient. In a sense this chapter is about the slow and deliberate attempt by the tumor to evade the T cell immune system and cause the death of the patient. This is why the first few events that occur during the development of a tumor are some of the most important in a person's life.

One of the most important aspects of normal cells becoming cancerous is the development of DNA mutations. Normal DNA produces normal cellular proteins that end up as part of the cell's outer membrane. The outer membrane is exposed to the outside environment, which is inspected by the surveillance arm of the immune system. Cells of the immune system such as NK cells, macrophages, and some T cells come into regular contact with just about every cell in our bodies. If they don't have immediate access, these cells have access to areas where they would quickly assemble and follow the "breadcrumb" trails to the cancerous cell and respond to it. In this case, the breadcrumbs would be a wide variety of altered proteins secreted by tumor cells and any tumor-associated debris left over from an NK cell attack.

This brings us back to the issue of DNA mutations occurring in a normal cell before it becomes cancerous. When the DNA breaks occur, the repair mechanisms of the cell mend them. Sometimes the repair is not correctly done. When DNA affected this way expresses a cell surface protein, the amino acid sequence of that protein is also affected and is often incorrect. The incorrect proteins begin to accumulate on the cell surface when all of a sudden they are detected by the immune surveillance system. Some of the "altered" proteins are also secreted and leave a trail for macrophages and T cells. These proteins have been referred to as neoantigens. They are not quite tumor antigens and the cells they come from are not quite tumor cells because unregulated growth has not yet occurred (Robin et al. 1997). That event usually takes additional DNA mutational steps. Yet, in many cases the immune system has been alerted to the presence of these cells primarily by the action of NK cells, whose job is to kill a limited number of these cells, debris from which can be processed by the rest of the immune system.

Some of the neoantigens are recognized by the T cell receptor (TCR). T cells in the form of responding CD4 T cells accumulate at the site of the precancerous normal cells. CD4 T cells need the TCR-neoantigen contact and CD4 protein-to-MHC class II protein contact to respond. Often, a low level of MHC-class II is expressed by neoantigen-expressing cells in this state and a limited amount of CD4 T cell proliferation occurs. The anti-neoantigen CD4 T cells release small amounts of IL-2, IL-4, and GMCF. IL-4 and GMCF are needed to change monocyte/macrophages into mature DC. This kind of CD4 T cell response produces neither, and an immature DC develops. Immature DCs turn neoantigen-specific CD4 T cells into Treg cells, which prevent T cells in the vicinity of the precancerous cell from attacking it (Mahnke et al. 2007). This is a normal and vital type of T cell response to a cell that may or may not be a cancer. It gives the cell in question the time to either go back to normal, which happens especially in non-cancerous situations, or continue to become malignant. This normal function is used by the potentially cancerous cell to buy time.

Several things can happen in this case. The cell can go back to being normal because the DNA repair enzymes cut out the DNA causing the expression of the neoantigens. However, depending on the veracity of the next step, there may be no chance for the immune system to catch up to what is about to happen. We are back at the DNA mutational steps of the very same cell. At this point, there could be two separate outcomes depending on the nature of the mutations. First, as the DNA break/repair mutations accumulate, in some instances two things may happen at once. As soon as the genes that control the cell's growth are disrupted and the cell begins out of control division, almost simultaneously, a DNA gene promoter (Walther and Ulrike 2000) is spliced (by mistake or on purpose) in front of the genes that make the MHC class II proteins. This causes unusual amounts of the protein to be expressed on the surface of each of the now cancerous cells. The T cell immune system now has the ability to begin a strong CD4 T cell response. This outcome can often, but not always, lead to the destruction of cancer cells that have developed this way. The other outcome is not so good for the patient. The genes for MHC class I and class II are silenced through multiple mechanisms (Wright and Ting 2006). Needless to say, the cancer cell that develops almost instantly has the ability to move and grow anywhere in the body because the most lethal part of the immune system (T cell) cannot see it right away.

These two types of cancers are still in the developmental stages. Both can be detected by the T cell immune system or can evade destruction. The first tumor to look at is the one with increased expression of MHC class II proteins. As soon as uncontrolled cell growth and the up-regulated MHC class II proteins occur, true tumor antigens are also created. These are proteins, recognized by the "innate" T cells. These T cells are programmed to respond to peptide-protein configurations that are passed along to the next generation. The peptides on these new tumor cells initiate a strong anti-cancer T cell response. If the response occurs rapidly, it can kill every tumor cell. That's the best-case scenario, related to "spontaneous remission". If the tumor antigens form later, that leaves only high levels of MHC class II protein on the tumor cell surface. The tumor cells can adversely affect the anti-cancer T cell response in two ways.

Both situations have to do with the interaction of T cells with just the MHC class II protein on the surface of the tumor cell. CD4 protein on the cell surface has an affinity for MHC class II. T cells have another ligand protein derived from the lymphocyte activation gene-3, or LAG-3 (Liang et al. 2008). Its receptor is also the MHC class II protein. The two T cell proteins (CD4 and LAG-3) attach themselves to the tumor MHC class II protein and cause the same cross-linking events that lead to T cell inhibition. Many of these intercellular events (inhibitory pathways) induced within the tumor cell can lead it to secrete immunosuppressive lymphokines such as IL-10 and TGF-B (Liu et al. 2007). These lymphokines can induce or convert normal or anti-tumor T cells into Tregs. Now you have Tregs being created before the expression of T cell-recognizable tumor antigen. This is not a good situation for the T cell immune system. The T cells attached to the MHC class II "only" expressing tumor cells then develop into suppressive T cells because cross-linking of both LAG-3 and CD4 together or separately has been shown to produce unresponsiveness in such T cells (Liang et al. 2008). So again the tumor creates T cells that cannot respond to infections.

Tumor cells that do not express MHC proteins can adversely affect any attempt of T cells to attack the tumor. However, the tumor cells can be seen by the T cell immune system even if they are almost invisible because of the lack of MHC proteins. The T cell support comes into play here. Natural killer cells can attack any cancer because the cancer cell membrane is abnormal. But their numbers are limited unless they are summoned to the site. When macrophages digest the debris from a tumor killed by NK cells, they process the tumor proteins and present them (on the macrophage cell surface) to CD4 T cells in conjunction with self-MHC (see Chapter 2 on antigen presentation). The CD4 T cells that have the receptor for these cancer antigens begin to proliferate and release the DC generating lymphokines as well as IL-2, which is needed by CD8 T cells to become anti-tumor killer cells. Mature DCs are immediately formed and begin to present even more tumor antigens in a much more efficient manner. Some of the activated DCs leave the cancer site and go to the draining lymph nodes. They release a small army of angry (activated) NK cells, which cut a direct path to the tumor. The DCs also activate CD4 helper and CD8 killer T cells, which too will eventually be en route to the tumor site.

In the meantime, the macrophages and DCs continue to activate CD4 helper T cells at the tumor site. One of the many lymphokines CD4 T cells release in these situations is gamma interferon (INF-g), which can enter a cell and activate the secretion of MHC class I and class II proteins. In this case INF-g enters the tumor cells and makes them express these proteins, enabling CD4 and CD8 T cells to proliferate and attack the tumor (Guerry et al. 1987). All of this takes time and unfortunately the cancer is meanwhile moving about the body because of its initial freedom. However, the reaction just mentioned can work at a distance because it is based on soluble small immune proteins (lymphokines). The lymphokines cannot work systemically (throughout the body) unless there is a widespread immunogen, such as a virus. Therefore, this immune cascade can work in a local area, such as a particular organ or tissue. Since these types of almost invisible tumors quickly become fully developed, they devise ways to fight back. They actually can use this type of T cell immune response to retard the formation of a formidable attack by the anti-cancer T cell immune system.

How an Established Cancer Influences the Anti-cancer T cell Immune Response

Established tumors are different from the ones mentioned above for several reasons. First, they have been through the initial attempts by the T cell immune system to destroy them. In some respects this has shaped or altered the population of tumor cells to allow them to continue to assault the immune system. Some of the tumors try to cripple or disable the T cell immune system. Second, established cancers are very different from the initial cancers because of their DNA mutation rate. The mutation rate of a tumor is important because it dictates the direction and effectiveness of the anti-cancer T cell response.

Let's look at the situation of a solid tumor. A mass of cancer cells sits in the liver, skin, or intestine and multiplies. Because tumors can induce angiogenesis (formation of a blood supply within the tumor mass), they secrete proteins. Scientists have developed tests that can pick up these proteins and diagnose a cancer. The proteins include tumor peptide antigens that T cells have a receptor for. When the T cells get close to the tumor, the antigens try to bind to the receptor prior to contact with the target tumor cell. They actually block the ability of the T cell to make contact with the cell-bound form and allow the T cell to make contact with only the tumor-expressed MHC proteins. This can induce T cells to become anergic—not responding to tumors and pathogens. Since the TCR can see two distinct peptide antigens, it can cause immunological problems not related to the tumor. The same thing happens when a chain of tumor peptides is linked to the TCR. A TCR signal without the corresponding CD4/CD8-MHC signal results in T cell paralysis, or anergy (Edwards and Evavold 2010).

Let's go back to the shedding of proteins by the tumor cell. The tumor cells shed immunogenic tumor-associated proteins as well as numerous proteins of all kinds. What happens next is another assault on the anti-cancer T cell immune system. We have said that macrophages eat everything, cleaning up the mess other cells leave behind. When tumor cells are killed, macrophages eat their cellular debris. Once gorged with proteins, they can no longer function as traditional macrophages. Their membranes fuse together and they form a mass called granuloma (Warrell 2005). The granuloma is a normal biological phenomenon because it is supposed to isolate the material the macrophages have eaten. The mass itself takes on a tumor-like appearance. Now you have two masses in a small space, which creates enormous pressure on the normal cells surrounding them. This event triggers a series of inflammation-related changes. Now another gene is triggered in both the granuloma and the tumor. The gene produces a protein that induces angiogenesis—new capillaries are formed to feed the mass, creating a blood supply that provides much-needed nutrients for the survival of the granuloma and, unfortunately, the growth of the tumor. The new blood supply provides an avenue of escape for the tumor cells (Kirsch and Black 2004).

The increased mutation rate of established cancers causes another problem that directly affects the anti-cancer T cell response. Normally almost all cells have the same genes expressed by T cells except that they are silenced during embryonic stem cell differentiation. However, when these cells become cancerous they get de-evolved into a stem cell–like state. DNA promoter genes, which are part of the human genome, are placed in front of genes that are not normally expressed by normal cells during the formation of a cancer. In some cases, some of the promoters are placed in front of lymphokine genes inside tumor cells, which are then activated to produce their lymphokine proteins. Some of these lymphokines, which are also called cytokines, can have a devastating effect on the anti-cancer T cell response. The cytokine proteins tend to induce T cell suppression, or anergy. This is bad for the entire biological system. The system-wide effects occur because there are millions of cancer cells throughout the body, secreting immunosuppressive lymphokines.

Let's look at what happens in a tumor mass when this occurs. The first protein secreted from cancer cells is T cell growth factor-beta, or

TGF-B (Liu et al. 2007). TGF-B is a cytokine that has a complex role in the immune system. It regulates lymphocyte proliferation and the ability of T cells to differentiate into many sub-populations with specific functions. It inhibits CD4 T cells from differentiating into helper type 1 T cells, or "first responders" of the anti-cancer T cell immune response. TGF-B also has a negative effect on DCs. It inhibits their ability to secrete IL-12, the lymphokine needed for naïve CD4 T cells to become type 1 helper T cells. TGF-B also makes DCs secrete IL-10, which inhibits DC and T cell function. In conjunction with TGF-B, IL-10 inhibits Tregs (Jonuleit et al. 2000). So you have the tumor mass secreting this cytokine where there is expression of both MHC proteins and multiple tumor-associated antigens. T cells with anti-tumor receptors are attracted there. But instead of getting killed, or becoming anergic, they get converted into Tregs, which can stop an anti-tumor T cell response in its tracks.

Tumor cells in this situation secrete TGF-B and can induce DCs to secrete this cytokine through a mechanism that is yet unknown. When the DCs are incubated with certain tumor cells DC activation is inhibited and TGF-B is elaborated (Dumitriu et al. 2009). This is a bad situation for the anti-cancer T cell immune response. Nevertheless, anti-tumor T cells that arrive in a tumor site after being activated by DCs in the lymph nodes do not get to attack the tumor, but are instead transformed into Tregs. The transformation is the result of cytokines entering the cell and activating several genes. One of the genes is the foxhead box P3 or FOXP3 gene (Walker et al. 2003). It's a repressor gene, which means it suppresses other genes of that T cell. The genes that get suppressed are associated with the ability of the T cell to kill its target, such as genes for perforin release used by CTL to bore a hole in the target cell (Chapter 1). This can happen gradually because Tregs can kill DCs presenting tumor peptides and MHC. The T cell that is to become a Treg still has the receptor for the tumor, but it keeps that T cell (now Treg) close to the tumor. So this Treg makes multiple contacts with the tumor, but does not affect it. If any other T cells approach the tumor, the Treg inhibits them, either by cell-to-cell contact or by secretion of lymphokines, in particular IL-10, which by itself inhibits T cell activation.

Unfortunately, there is something special about the tumor-specific Tregs. These cells are very aggressive. They can leave that tumor site, find another one, inhibit the anti-cancer T cell response there, and move on to the next distant tumor mass. As long as there is a blood supply, these Tregs can travel from one tumor site to another. This is just the tip of the iceberg (Mougiakakos et al. 2010). Tumor Tregs are capable of doing much more damage to the anti-cancer T cell immune response than previously imagined. These cells can easily move through the bloodstream, partly because they used to be anti-tumor effector T cells (Wang et al. 2005), which can go wherever they want. The tumor Tregs can also move over surfaces. A part of what I did as a cancer researcher was to examine the T cell immune response to human melanoma in the lab (*in vitro*). I would usually incubate a patient's T cells with their tumor and examine the results of that interaction (Alexander et al. 1989). Sometimes it would be T cell proliferation (Guerry et al. 1987) and sometimes the generation of cytotoxic T cells (CTL). The latter assay required me to count the T cells that were incubated with the tumor prior to their addition to the CTL assay. One day I was counting a sample of melanoma-immune T cells and someone interrupted me. When I turned back to look at the slide, several of the cells had moved quite a distance When I watched the slides, say, for 45 seconds to a minute continuously, I noticed extreme movement of some of the T cells. These were undoubtedly activated T cells, but it was impossible to tell of which phenotype. This is what has been described in articles about Tregs and their incredible mobility (Matsushima and Takashima 2010).

This is an important aspect of the anti-cancer T cell response because tumor-specific Tregs have an affinity for DCs. The situation is another complication for the anti-cancer T cell immune system. Since Tregs move so fast, they find the DCs first; often, even before naïve tumor-specific T cells do. This is a problem because of what the Tregs do to the DCs. First of all, they have two proteins on their surface that immediately allow them to anchor themselves to the DC surface. The first of these proteins is called LFA-1 and it attaches itself to a protein on the DC surface called ICAM-1 (Onishi et al. 2008). The second is called LAG-3 and attaches itself to the MHC class II protein on the DC (Liang et al. 2008). The first interaction with ICAM-1 is the initial contact and it cements the two cells together. The second one is the problem. This interaction between LAG-3 and MHC class II has two consequences. First, it ties up the MHC class II protein and makes it unavailable to naïve anti-tumor T cells, which need this contact to become activated and attack the tumor. Second, and much worse, the DCs begin to de-differentiate. This means they reduce production of surface proteins (CD80 and CD86) required for anti-cancer T cell activation. Additionally, the DCs produce Treg-inducing lymphokines (TGF-B and IL-10) and the attached Tregs re-program the DCs to go to the nearest lymph node (draining). This lymph node is where activated DCs have gone to arm T cells against the tumor (Bayry et al. 2007). Now you have this influx of Treg-inducing DCs converting the "just activated" anti-cancer T cells into more Tregs. This is not a good situation for the anti-cancer T cell immune system to face, but it is even worse when immunotherapeutic treatments, which are based on DC immunology, are used to fight the cancer.

Established tumors can use cancer-induced autoimmunity to divert the attention of the anti-cancer T cell immune system. To a certain extent the tumor population sacrifices some of its members to induce mutations in the TCR. This happens when the tumor is highly visible to the T cell immune system. There is a period when the T cell immune system attacks the tumor without inhibition. This results in the death of many cancer cells, but it also induces hyper-stimulation of T cells that have the TCR that recognizes the tumor antigens this cancer is expressing. The T cell immune system partly works through mutation of the TCR. Innate or natural anti-cancer T cells have receptors already expressed that attach to a tumor cell and either proliferate (CD4 T cells), releasing instructional lymphokines, or kill the tumor (CD8 T cells). Both these activities require that the T cell divide multiple times. Now there is the adaptive T cell immune system that relies on the ability of the TCR to mutate and conform to the threat, which in this case is a tumor peptide. Massive cell division is required in this case. This part of the T cell immune system works best with viral threats because there is a system-wide stimulus that induces this type of hyper T cell proliferation. Eventually the TCR mutates and conforms to the threatening antigen. However, tumor cells have been able to block the initiation of this process by the activation of STAT3, which is an oncogenic signaling pathway. The pathway blocks the ability of tumor cells to secrete proinflammatory mediators that initiate T cell proliferation necessary to induce an adaptive anti-tumor T cell response (Wang et al. 2004).

So these are the conditions that lead to cancer-induced autoimmunity. As the anti-cancer T cells respond to the tumor, the massive amount of proliferation causes the TCR to mutate. The mutation in the receptor is usually minimal because this is not like a systemic viral infection, but local, at one or several cancer sites. The mutation causes the T cell to recognize a normal cell. Usually this cell is related to the tissue-type of the cancer, in other words, the normal cells the cancer developed from. Unfortunately, this is a hallmark of many different forms of cancer. Many cancer patients have been diagnosed with cancer because of autoimmune reactions caused by the T cell response to a developing or established cancer. Many of these symptoms are related to the tissue-type of the cancer. For instance, in some patients with melanoma, the first symptom of a tumor is either peripheral pain from nerve damage (Visovsky 2005) or vitiligo (Gui et al. 2007). Melanoma develops from melanocytes and these cells are a part of what is called the "neural crest". Melanomas are derived embryonically from neural tissue. Depending upon what tissue the cancer develops from, there is often a corresponding attack of normal cells.

Clinically, this autoimmune attack on normal cells is a positive sign because it is an indicator of a brisk T cell immune response to the growing tumor (Fujimoto et al. 2006). Despite this, the attacks are dangerous for the patient because the destruction of normal tissue further diminishes the patient's health. This situation has been created by the tumor because of its recognition by the immune system. The system reacts to this phenomenon by producing Treg cells, whose function is to stop reactions like this. Unfortunately, the cells that get recruited or reprogrammed to do this job come directly from the anti-cancer T cell pool (Liu X et al. 2009a). This means the number of anti-cancer T cells is diminished and the pressure on the tumor is lessened. In this way the tumor has diverted the attack on itself. Established tumors do this very efficiently. It's a numbers game. The tumor gives up millions of cells to the immune system in exchange for more time to grow and cause the death of the patient. These issues also come into play when cancer is treated by new immune-based treatments. Knowledge of the mechanisms the tumor uses against the immune system will help scientists to successfully treat these cancers.

Interference in Immune-based Cancer Treatment by Cancer Cell Influence over the Anti-cancer T cell Immune System

In recent years scientists realized that our T cell immune system was fully equipped to destroy a growing tumor and leave behind a T cell memory, so that if the same kind of tumor reappeared, it too would be eliminated. As we knew more about the T cell immune system it became clear that there was a way to manipulate cells of the immune system so as to make them attack a tumor that previously was resistant to traditional anti-cancer treatments such as chemotherapy. It was apparent that the T cell immune system was capable of stopping a tumor, but there were elements that prevented that from happening. Scientists first wanted to bolster the T cell immune attack through immunotherapy, by isolating tumor-infiltrating T cells (TIL) from surgically removed tumors and expanding them in the lab using the lymphokine IL-2. After the cells were expanded they were infused back into the patient to attack the remaining cancer cells (Rosenberg et al. 1988). This treatment resulted in the regression of some tumors in a small number of patients. The treatments induced measurable (laboratory) responses in a majority of the patients, but because the initial clinical trials were administered to late-stage patients, many of them died despite the positive responses

Scientists realized that this technique was fraught with problems. First of all, once TILs were removed from the body and grown in the laboratory, changes took place within the cells. The most important changes had to do with adhesion proteins expressed on the surface of the cell (Krummel and Macara 2006). These proteins are used by the T cell to move through the body, adhere to the tumor and stay put. The culturing process disrupts the expression of these proteins and when they are put back into the patient they have a hard time making it to the tumor site. Many times these infused TILs would end up in the lungs, which would cause problems for the patients because the cells would induce granulomas. This could cause an embolism, a life-threatening condition. Another complication was that TILs not only contain anti-cancer T cells, but also Tregs. Tregs are an essential component of the anti-cancer T cell population in a tumor. Fortunately, Tregs do not multiply in the laboratory unless cultured with mature DC (Yamazaki et al. 2003). Still, the presence of Tregs in the expanded population of T cells infused into patients was a problem. An even greater problem was the presence of Tregs already in the patient prior to infusion with expanded TILs. Fortunately, scientists found a way to deal with that.

The procedure is called lymphoablation, or removal of all the lymphocytes of the patient prior to infusion of the expanded TILs. This removes tumor Tregs from the patient's body. The method uses chemotherapy and radiation to kill circulating lymphocytes. These cells are replenished by the bone marrow, where lymphocytes originate. Shortly after radiation and chemotherapy, however, the patient's immune system does not exist and that leaves them vulnerable to infections. This procedure did increase the objective response rate from 60% to over 70% (Dudley et al. 2008). Still, the therapy resulted in complete regression of the tumor in only a minority of patients treated. When radiation induces death of cells, the cells most vulnerable to ionizing radiation are those that divide frequently. For T cells there is no problem here because they are made to divide and reproduce. It is part of their make-up.

However, the myeloid series of cells do not reproduce this way. They are born in the bone marrow and then change as the need arises, but generally they do not divide. Ionizing radiation cross-links the DNA chains and the cell dies. For cells that do not divide as part of their function the radiation has a minimal effect. The monocytes, macrophages, and DCs are basically radio-resistant. They stay put after the treatments and newly expanded anti-cancer T cells are introduced into the patient's body. The DCs, which have been found by the Tregs contained in the immunotherapeutic T cell infusion, begin to make even more Tregs that stop or limit the effectiveness of the therapeutic TILs. The cancer cells set up this situation. Dendritic cells were in the tumor mass, converting anti-cancer T cells into Tregs, and all of a sudden all the T cells died, but the DCs remained. When the new therapeutically expanded anti-cancer T cells arrived, the DCs continued to do what they were doing before they got bombarded with radiation. Scientists recognized this and started to think about what was actually happening. The myeloid-lineage cells were actually directing the anti-cancer T cell immune response. So the next step was not much of an intellectual leap because the evidence from these early clinical trials sent a clear message. Maybe we should be concentrating on another type of cell for immunotherapy.

Scientists chose the DCs because of their heightened ability to present tumor antigens and because the cells could activate CD4 type 1 anti-cancer T cells in the lab and in mice. These T cells are responsible for initiating the kind of T cell response that eliminates the tumor and leaves memory T cell behind to deal with tumor cells that escape the initial attack. The immunobiology of DCs had been studied for years before they were chosen to be the newest weapon in immunotherapeutic treatment of cancer. The initial experiments involved removing two things from the patient: blood and a biopsied sample of the tumor. Monocytes were isolated from the blood and cultured with IL-4 and GMCF to induce them to become mature DCs. Usually the tumor sample was turned into cellular debris by either repeated cycles of freezing-thawing or the use of a proteolytic enzyme, which cut the proteins into small pieces. Once the DCs were mature the cellular debris from the patient's tumor was added. Now the DCs would process the tumor antigens and then were prepared to present the tumor antigens to the patient's naïve anti-tumor CD4 T cells, which would begin the anti-cancer T cell response (Redman et al. 2008).

The initial trials of a "DC cellular vaccine" were puzzling to scientists and doctors because all indications were that this procedure should have induced cancer regression, but did not. As with the TIL trials, there was evidence of a new anti-cancer T cell response in the lab using treated patient T cells, but still there was no regression of the tumors. That actually is not an absolute because a few of these treated patients developed stable disease and even fewer had some remissions, but in reality the therapies were a failure. The positive results with the OR rates were enough to continue with this line of clinical immunotherapeutic research.

There have been many attempts to get this type of DC-based therapy to initiate a positive anti-cancer T cell response that would lead to the destruction of the tumor. These clinical trials failed partly because of DC immunobiology. First of all, using DCs as a tool to induce an anti-cancer T cell response is not as simple as it would appear to be. The initial observation that DCs present tumor antigens more efficiently than any cell in the body is correct. However, this ignores the fact that DCs have a complicated relationship with all T cells, not just anti-cancer T cells. If we put aside the matter of cancer for a moment and think about what else the DCs do in our bodies, that may give us a clue as to why DC-based immunotherapeutic treatments might not be working as well as expected. Neoantigens (Robin et al. 1997, Katz and Bonavida 1989) are slightly altered proteins a cell produces under certain circumstances. I earlier used this term in reference to a normal cell that was being transformed into a cancerous cell, but the fact is that many cells in our bodies go through a similar stage for a multitude of reasons. In Chapter 10, I used the example of a person who was prescribed steroids for inflammation of the lungs. The steroid basically reduces the inflammation to the point where the immune system can cope with the problem. However, other cells are affected and among them are the islet cells in the pancreas. These cells produce insulin, which is vital to our ability to metabolize sugar. Without this protein (insulin) our bodies begin to decay.

What does this have to do with DCs and the control cancer cells have over the anti-cancer T cell response? Dendritic cells actually play an important role in this process. Steroids tend to affect islet cells in a way that is not good for their survival. Steroids alter the cell membrane and cause the cells to produce slightly altered cell membrane proteins. These proteins now are considered by the immune system to be questionable or neoantigen-like. A passing NK cell notices this and stops. It kills a limited number of these cells. In come the macrophages, which quickly ingest the debris. They process the antigens and present them to CD4 type 1 T cells. A limited number of CD4 T cells that have a receptor for this protein begin to proliferate and elaborate DC-inducing lymphokines. Monocytes in the area are converted into DCs, but because the signal was slight, the DCs are immature. Immature DCs usually induce Tregs. Tregs are created from the neoantigen-reactive CD4 T cells. The newly created islet-neoantigen–specific Tregs now protect the islet cells from attack mediated by other T cells (Long et al. 2009). This gives the islet cells time to return to normal because if the steroid treatment goes on too long and the neoantigens are expressed by every islet cell, a full-blown autoimmune attack ensues, which results in diabetes.

This is part of the relationship between a DC and a T cell. It is not related entirely to the anti-cancer T cell response, but in some respects it mirrors what happens during an anti-cancer T cell response. Using DCs to treat cancer can actually accelerate tumor growth because once immunotherapeutic DCs are infused into the patient they present antigens, but they come under the control of Tregs. These DCs can also instruct Tregs to exert even more control over the anti-cancer T cell response. Dendritic cells do this because they can actually facilitate Treg cell division. Tregs express CD25, which is the receptor for IL-2, but IL-2 alone supports only a limited amount of Treg proliferation. Unlike CD4 helper T cells, CD4 positive Tregs need to come in contact with the B7 protein on the DCs and a small quantity of IL-2. A normal CD4 helper T cell needs only IL-2 in order to proliferate, but CD4 Tregs are different in that regard. So DCs prepared for immunotherapy can actually generate Tregs almost immediately because they have been activated and their expression of B7 (CD80) is very high.

How Scientists and Clinicians Are Trying to Improve DC-based Immunotherapeutic Treatment

The problems outlined above are only a sample of those relating to the ineffectiveness of DC-based immunotherapy. Again, the idea that DCs are the best choice of cell to use in immunotherapeutic trials is sound because of the potential positive role these cells can play in the successful treatment of cancer. Once the complications arising from the relationship that DCs have with Tregs are addressed, it is hoped that these therapies can actually cure people of their cancers. The immunotherapeutic treatment of cancer to a certain extent is painless. If it is performed under the proper conditions, removes the cancer and leaves a memory behind, the future of cancer treatment will be changed forever. The root cause of some of the problems with DC immunotherapy lies within the DC itself. This is called "translational immunobiology" (Chapter 9), meaning that there are intercellular signals within the cell initiated by either cell-to-cell contact or contact between the DC and lymphokines/cytokines. These signals within the cell dictate the way DCs behave.

Let's start with the B7 protein expressed by the DC. There are certain types of antibodies that when attached to this protein on the DC surface induce the DCs to do something quite extraordinary. The DC begins to convert normal CD4 helper T cells and Tregs into an autoimmune effector T cell. These cells are stimulated by a lymphokine called IL-17. IL-17, secreted by the DC under these conditions, takes part in the transformation of the two types of T cells mentioned above. Th-17 T cells can attack normal tissues and also break self-tolerance. When islet cells are being protected by neoantigen-specific Tregs, the Th-17 T cells can override that protection. Once this happens, the islet cells are vulnerable to attack by either other neoantigen-specific effectors or by the TH-17 T cells themselves. They are thought to play a major role in the development of arthritis and multiple sclerosis. Obviously this is a bad thing, but if anti-tumor Tregs are converted into this T cell phenotype they revert to anti-cancer T cell effectors. Now they can recognize the tumor and kill it again. Treatment of mice with an antibody directed at surface proteins expressed by DC that mediate this conversion of tumor Tregs into effectors has demonstrated tumor regression (Radhakrishnan et al. 2008a). Now comes the most interesting part. Scientists discovered that an intercellular pathway (phosphorylation of syk and phospholipase-C) is activated during this process and its inhibition during this process induces DCs to secrete IL-12, which activates tumor-specific CD4 type 1 helper T cells (Radhakrishnan et al. 2008b). So, once the DCs are activated by this anti-B7 antibody, they can be treated with the inhibitor piceatannol and produce a lymphokine (IL-12) that potentially could result in the removal of the cancer. This type of treatment should increase the ability of DCs to jump-start an anti-cancer T cell response. Treatments like this one may be even more powerful than the reprogramming of anti-cancer Tregs to become anti-cancer effector T cells because IL-12 induces brand new anti-cancer T cells, which include CD8 killer T cells, the most lethal weapon of the immune system.

Another related pathway is induced by the activation of toll-like receptors expressed by DC. Activation through some of these receptors induce DCs to secrete IL-10 and TGF-B, which are essential in initiating this conversion process of normal naïve T cells and anti-tumor T cells into Tregs. The TLR family comprises receptors that recognize pathogen-related molecular patterns, and there are 10 of them. Five of these receptors (TLR 1, 2, 4, 5, and 10) recognize bacterial products, while the remaining three (TLR 3, 7, and 9) see viral nucleic acids. Some of the natural ligands of these receptors are lipopolysaccarides and unmethylated CpG DNA motifs (free-floating DNA from disintegrated bacteria). The DC expresses many TLRs, but ones that seem to be involved with IL-10 and TGF-B are TLR-4, TLR-2, and to some extent TLR-7/TLR-8. The ability of ligated TLRs to induce secretion of these two lymphokines by DC is thought to involve intercellular pathways dependent on p38 MAPK induction. Inhibitors of this kinase (p38) switch on IL-12 production by the DC. The DCs, however, must be activated through the TLRs and then inhibited by anti-MAPK inhibitors to initiate this type of CD4 helper type-1 activation mediated by IL-12 (Kalinski et al. 2010).

Another development with DC activation may hold promise to facilitate an anti-cancer T cell response. This has to do with a gene that produces a kinase (serine/threonine kinase) protein and is expressed by cells of the immune system such as T cells and DC. The protein product of this gene is called hematopoietic progenitor kinase 1, or HPK1. Initially it was determined that this protein regulated T cell functions through its ability to inhibit intercellular pathways leading to T cell activation such as NFkB pathway. Later, it was discovered that HPK1 controls the ability of DC to become activated. The protein product of this gene does not inhibit DC activation but limits it. Mice that had this gene removed had DCs that were 10 times as sensitive to being activated as their normal litter-mates. The DC from these HPK1 negative mice secreted vast amounts of IL-12, IL-4 and IL6, all of which are powerful inducers of CD4 type 1 helper T cells. In fact, the HPK1-negative DC could actually eliminate established tumors in normal mice (Alzabin et al. 2009). For the first time, apparently, DC activation could be increased without blocking intercellular pathways with chemical inhibitors. This is an exciting development with respect to DCs being used for immunotherapeutic treatment of cancer.

This is an experimental concept that has been demonstrated only in mouse models. It has not been applied to human cancer. However, the technology to silence the gene that controls DC activation is in place. It is called RNAi treatment. It effectively interferes with the expression of any given protein by a gene. First you must know the sequence of the gene and the resulting sequence of the messenger RNA that assembles the protein in question. RNAi means interfering RNA. This involves making a complementary copy of the RNA that makes the protein you want to block. This complementary copy binds to the real RNA that is supposed to make this protein. Once the copy binds to the real RNA, it cannot be used as a template to form the protein. This is a way to silence any gene product produced by any cell (Ross 2003). In humans, gene silencing could be performed on immunotherapeutic DCs that have been pulsed with tumor antigens from the biopsied patient's cancer. This would have to be done just before the DCs are infused into the patient. This ability to block this HPK1 gene product by RNAi inhibition is temporary. The bound HPK1 RNA eventually is discarded by the DCs and they begin to function (after some time) as normal DCs do. The HPK1-blocked DCs can enter the tumor mass and secrete vast amounts of these lymphokines, which can induce a positive anti-cancer T cell response.

The authors of this study discussed the cautionary aspects of such a treatment. Dendritic cells activated in this manner have the ability to induce autoimmune responses. These can be quite serious for the patient. However, some of these autoimmune reactions are indicators of a positive anti-cancer T cell response, but they have to be taken seriously because they can lead to the death of the patient. Patients must be monitored during a procedure like this one. They may need steroids to dampen any autoimmune reactions induced by it. The other precautionary warning about silencing the HPK1 gene is that once the gene is turned off, the DCs survive much longer than usual. This may not appear to be a problem, but if immunotherapeutic DCs do not go through the usual cycle of life and death they may become malignant. If a DC turns cancerous, it could cause death before the problem is diagnosed.

One more potentially risky therapeutic DC alteration has to do with the cell-to-cell contacts between DC and Tregs. Tregs have on their surface a ligand called LFA-1. The receptor for this ligand (on DCs) is called ICAM-1. The ICAM series of protein receptors are adhesion molecules. The receptors are important when cells make contact with other cells. As a matter of fact, some cells such as human granulocytes (neutrophils and eosinophils) depend on this surface protein to adhere to endothelial cells of blood vessels. In this case they are called carcinoembryonic antigens (Beauchemin et al. 1999). The interaction between this ligand and receptor are immediate. The cells are basically cemented together. Once this happens, the Tregs begin to exert control over the DCs using all the cell-to-cell contact receptors they have. There is one that controls DCs in such a way as to basically de-differentiate the cells. However, before mentioning that, it is important to realize that this interaction between LFA-1 and the ICAM-1 molecule on the DCs should be temporarily inhibited by using an anti-LFA-1 antibody because without it the door is opened for the Tregs to dominate DCs. You do not want this to happen if you are using tumor antigen–pulsed DC to treat cancer.

Another interaction between Tregs and DCs causes DCs to be non-responsive to activation signals. Activated T cells and particularly Tregs express a gene called lymphocyte activation gene-3, or LAG-3 (CD223). Biologically this protein is a member of the CD4 family of ligands. The receptor for these proteins is the MHC class II protein expressed by DCs. When these two proteins interact, two things happen to the DCs that diminish their ability to induce a positive anti-cancer T cell response. One, the MHC class II protein is tied up. The DC needs this protein to present tumor antigens. To have it occupied by a Treg cell inhibits the DC's function. Two, this interaction between LAG-3 and the MHC class II protein induces a negative signal within the DC. That in turn activates a negative pathway called ITAM (Liang et al. 2008). When this pathway is activated, the DC lowers the expression of B7.1 (CD80) and B7.2 (CD86). These two proteins are essential for the DC to activate anti-cancer TH-1 CD4 T cells. It turns out that Tregs created from anti-cancer T cell effectors (tumor-specific Tregs) have the ability to find, bind, and inhibit DCs at a very high rate (Valitutti and Dupre 2010).

Initially it would be good to inhibit this interaction by using an antibody against the MHC-class II protein expressed by the DC. However, that would be self-defeating because the DC needs unfettered access to this protein by anti-cancer CD4 T cells. Fortunately, a series of intercellular pathway inhibitors can block the activation of this negative ITAM pathway. Methyl-B-cyclodextrin effectively inhibits the pathway (Cifuentes et al. 2008). Using an inhibitor like this one may give the immunotherapeutic DC time to activate anti-cancer T cells without interference from Tregs.

In conclusion, cancer cells influence the anti-cancer T cell response on many levels. Understanding this influence, which amounts to interference, deception, and immunosuppression, can lead to effective treatments for cancer. Tumor cells by their very nature challenge the T cell immune system in ways that could potentially lead to the death of the patient. These challenges must be thwarted to control and contain cancer. In many cases, once a cancer is established, it is likely to kill the patient. However, some cancers like forms of kidney and prostate cancer grow so slowly that the patient may die of old age. It is the young cancer patients who die that are of major concern. Over the last 20 or 30 years there has been great progress in fighting cancer. Some cancers that were deadly are now treatable. The cancers that remain are largely untreatable because they have a devastating effect on the only system equipped to fight them. This is the anti-cancer T cell immune system. Although the T cell immune system is prepared to fight cancer, it cannot under certain circumstances. These circumstances, induced by the cancer itself, have to be controlled by outside intervention.

Immunotherapeutic treatment of cancer clearly has advantages over chemotherapy and other traditional treatments. Chemotherapeutic treatments for cancer have also undergone significant changes over the years. These changes involve the ability of the chemical agents to target specific pathways in tumor cells that allow these cells to grow and interfere with the function of the T cell immune system and the entire biological system of the body. One of the major advantages of successful immunotherapeutic treatment of cancer is the development of a T cell-mediated immunological memory. This memory of the cancer, after it has been eliminated, is a natural result of the T cell immune system functioning in a proper manner. Many aspects of the generation of anti-cancer memory T cells are not quite understood (Obar and Lefranois 2010). Nonetheless, the proper formation of this memory should guarantee that the tumor will never return. This is something that many chemotherapeutic agents cannot do.

With this continuous generation of knowledge concerning the ability of a cancer to influence the anti-cancer T cell response, it should be possible to cure this disease, given enough time. Scientists, clinicians, and laboratory workers worldwide are battling cancer. Within 20 years, we may be able to successfully treat well over half of the diagnosed cases of serious life-threatening cancers. Our bodies are intricate biological machines and the mechanisms that govern the way our T cell immune system functions are complex. Because of advances in laboratory technology, scientists have discovered more about the T cell immune system in the last 50 years than in the preceding 100. This knowledge is so vast that it takes years to make the proper interconnections that lead to new anti-cancer therapies. Now, because of the coalescence of some of this knowledge, new anti-cancer immunotherapeutic treatments are being developed. These new T cell immune-based treatments can potentially accelerate the eradication of cancer to the point where even the most difficult types of cancer will be reduced in size and eventually cured. They will include procedures that circumvent the influence cancer cells exert over the T cell immune system. This would allow anti-cancer immunotherapeutic treatments time to eradicate the tumor. When that day comes we will be one step closer to curing most forms of cancer.

Chapter 14

The T cell Anti-cancer Response, T cell Regulation, and Immunotherapy of Neural Crest Cancers

The neural crest consists of the ectodermally derived cells along the outer surface of the neural tube in the early stages of embryonic development. These cells give rise to spinal, cranial, and enteric cells, sympathetic ganglia, pigment cells, Schwann cells, and cells of the adrenal medulla. This is a standard medical definition of the neural crest tissues that give rise to cells of our nervous system. The only cell that seems out of place is the pigment-producing cell. There are over 15 different types of cancers associated with these tissues. The vast majority are brain cancers and only one is associated with the skin. Very few of the cancers that develop in the brain are benign. These include the neurofibroid cancers such as neurofibroma, which is derived from Schwann cells. Removal of this cancer almost always leads to complete recovery. Most brain cancers are malignant to one degree or another.

There are two major types of malignant brain cancers—blastomas and gliomas. They contain multiple subtypes. The blastomas originating from the sympathetic nervous system are quite lethal. They have the ability to move (metastasize) quickly to other parts of the body, making treatment difficult. Neuroblastoma is one example. The gliomas too contain several subtypes and are just as deadly. Glioblastoma multiforme is perhaps the most deadly example. This cancer is made up of a mixture of mature and immature tumor cells, some normal brain cells (astrocytes), and cells of the immune system. The presence of cells from the immune system is important and will be discussed at length later.

The next type of cancer occurs in the skin and is just as deadly as brain cancer. The normal cell in which this cancer develops is the melanocyte, which is responsible for skin pigmentation. The cancer is called melanoma. Melanoma has just about every characteristic of all the cancers that affect humans. There are benign forms as well as malignant forms. Because of its visibility in the skin, melanoma has probably become one of the most frequently studied cancers (Turkington 2005).

This discussion of cancers of the neural crest will focus on glioblastoma multiforme and melanoma. They have elements that typify the characteristics of all the neural crest cancers. They share proteins recognized by the T cell immune system as well as express unique proteins that are also recognized. We give a general description of these two malignancies and then describe their relationship to the T cell immune system. Also, we examine the problems the T cell immune system encounters when trying to fight these cancers. In order to further understand the relationship these cancers have with the T cell immune system, we look at the similarities and differences in the T cell response as they relate to patient survival. We address the question why these cancers are likely to respond to immunotherapeutic treatments. Then we review experimental immunotherapeutic approaches to the treatment of these cancers.

Two Representative Neural Crest Cancers: Glioblastoma multiforme and Melanoma

Glioblastoma multiforme

Glioblastoma multiforme (GBM) is a complex mixture of malignant and non-malignant brain cells and cells of the immune system. It is often detected in the cerebrum or spinal cord. The symptoms are headaches and disorientation, and it generally affects middle-aged people. Glioblastoma multiforme is the fastest-growing and most invasive of all the glioma cancers. It accounts for about 30 percent of the primary brain cancers. It has the ability to invade healthy brain tissue and is usually fatal (Ohgaki and Kleihues 2005). The median survival rate of a patient diagnosed with this tumor is one year and usually 4–6 months after surgical removal. Less than 5 percent of patients survive after five years from the point of diagnosis. Patients now live slightly longer because of the use of new methods in chemotherapy, surgery, and radiotherapy. However, the long-term survival (five years after diagnosis), despite improvements in treatment, has not significantly changed since the 1950s (McLendon and Halperin 2003).

Histological examination of GBM tumors has revealed the presence of a mononuclear cell infiltrate. This infiltrate comprises mostly T cells and some cells of the monocyte-macrophage lineage, suggesting an immune response is occurring with the tumor mass itself (Brooks 1978). The tumors have been shown to induce autologous T cell responses *in vitro* and *in vivo*. *In vitro* evidence suggests that the response results in the generation of cytotoxic T cells (CTL) capable of killing autologous tumor cells as well as histocompatibility-matched GBM cell lines (Tang et al. 2005). The *in vivo* response can be seen in histological sections of the tumor mass itself. T cells can be observed secreting gamma-interferon, which is identified with CD8 CTL activity. However, despite the presence of an immune T cell infiltrate, this cancer is uniformly fatal to the host.

The response by anti-GBM T cells is multifaceted in that it contains subpopulations of T cells that actually impede a productive (agonistic) immune response to the tumor. These T cells induce tumor-specific immune suppression (antagonistic). Evidence exists that CD4 T regulatory or Treg (CD25+, Fox3+) cells are present in the infiltrate that secrete inhibitory lymphokines such as IL-10 and TGF-B (Curtin et al. 2008). In addition, there are other T cells secreting these types of lymphokines that are induced to do so by their response to altered target peptides (Ruppert et al. 1993). These lymphokines can induce immune paralysis that may cause anti-GBM T cell responses to shut down. Even if T cells have the ability to kill GBM, these same cells may also interfere with the ongoing immune response at the same time because they are converting to the Treg phenotype (Curiel 2007). The fact that GBM-specific CTL activity can be measured is not by itself the best indicator of a response that would eliminate the tumor and leave behind memory T cells.

Another problematic aspect of this type of tumor is its location. It has been shown that the blood-brain barrier (BBB) can impede efforts to use chemotherapeutic agents to slow tumor growth or kill the tumor. More important is that it can filter or select which immune cells enter the brain (Anirban et al. 2007). Because of this, special attention must be given to this gateway to the brain. Permeability of the BBB is controlled by several factors, and multiple cells participate in this important process. T cells control most of this activity. They have the ability to secrete lymphokines that affect microglial macrophages and DCs. The microglial cells and DCs secrete chemokines that, in effect, cause the barrier to become resistant to signals that would increase its permeability. This retards the ability of immune T cells to attack the tumor. The T cells that initiate this cascade are thought to be Tregs (Phares et al. 2006). It is not clear why they are drawn to the tumor site. They must be controlled for a true anti-cancer T cell response to occur.

Melanoma

Melanoma develops in the melanocyte pigment-producing cell. Melanomas may metastasize to other parts of the body and are the most malignant of all the skin cancers. They occur predominantly in white women and men, usually over the age of 40, who have light skin and many moles. However, people of color do get melanomas, usually on the palms of the hands and soles of the feet. There is a greater incidence of melanoma in people who have dysplastic nevi, which are complex (irregularly shaped) moles thought to be the first stages of melanoma. There are four major types of melanoma: superficial spreading melanoma, nodular melanoma, acral lentiginous melanoma, and lentigo maligna melanoma. Intraocular melanoma is a rare type of melanoma affecting the eye (Turkington 2005).

To aid in the determination of the prognosis of melanoma, a staging technique was developed by Dr. Wallace Clark, Jr., of the University of Pennsylvania. It is called Clark's classification system (Clark et al. 1986). It classifies melanomas by how deeply they penetrate the layers of the skin and very accurately predicts the lethality of a melanoma lesion that appears on the skin. This development was an invaluable tool because histological sections of melanomas not only revealed the depth of invasion, but also showed the presence of a lymphocytic infiltrate (Clark et al. 1984). There was an extreme correlation between the presence of this lymphocyte infiltrate and survival (Kornstein et al. 1983).

Scientists were able to determine not only the type of lymphocyte (T cell) infiltrating the cancer mass, but also the method the T cells used to attack the cancer cells. It turns out that the T cells were responding to formerly melanocyte-specific proteins that were altered during the change from normal cell (melanocyte) to cancer cell (melanoma) and now recognized as foreign and called melanoma-associated antigens (MAA). CD4 T cells must see foreign proteins along with a self-protein called HLA-DR. As melanocytes make the transition from normal to dysplastic and then to cancerous, they start to express the HLA-DR proteins (Guerry et al. 1984). This, in combination with the MAAs, induced a CD4 T cell immune response that includes the formation of CD8 killer cells (Guerry et al. 1987).

Both these T cells (CD4 and CD8) could be visualized in sections of the tumor mass. The killer cells could be detected because they secreted gamma-interferon when they attacked the melanoma cells. This response seen in the skin was reproduced in the laboratory using melanoma cells cultured in the lab and purified T cells from the patients the cultured melanoma cells came from (Guerry et al. 1987). Other studies in the laboratory demonstrated that primary melanoma cells, but not metastatic melanoma cells, were able to present melanoma-specific proteins and other proteins such as tetanus toxoid. This was a significant finding because it showed that early malignant cancer cells were able to alert the immune system to their presence by acting like normal antigen-presenting cells. These and other *in vitro* studies were the framework and model for melanoma and many other types of cancer (Alexander et al. 1989).

How Similarities and Differences in T cell Responses to GBM and Melanoma Relate to Patient Survival

Early melanoma lesions were typified by the presence of an active T cell immune response. The presence of this response indicated a positive prognosis for the patient. However, this is not the case for GBM, where there is often a T cell infiltrate, but it does not correlate with the survival of the patient. This is the single most glaring difference between the immune responses to these two types of cancers. However, there is another difference in the responses that may reveal why GBM is fatal despite a T cell infiltrate. There is evidence that may shed some light on this disparity in the immune responses to these cancers and it comes from cancer-induced autoimmunity.

As melanoma cancers develop, they express the MAAs that induce a strong T cell response. Anti-cancer T cell responses are sometimes viewed by the immune system as anti-self autoimmune responses. This is where Tregs are recruited to infiltrate the cancer mass and they begin to shut down the anti-cancer T cell response as if an autoimmune reaction was in progress (Curiel 2007). Apparently this happens in GBM as well (Curtin et al. 2008). However, the T cell response in a melanoma tumor mass is able to remove the cancer cells, but in GBM it is not. There is some evidence suggesting that the type of Treg cell involved in the two immune responses is the cause of this very important difference in their outcomes.

There are two types of Tregs (Bluestone et al. 2003). The "natural" Treg is selected in the thymus as a part of the natural creation of T cells and is thought to contribute to homeostasis of the T cell immune system. The "adaptive" Treg is created from existing T cells under inflammatory conditions (Curiel 2007). It appears that it can be recruited to become anti-cancer T cells because the removal of these cells in GBM model systems can actually cause an increase in cancer size (Curtin et al. 2008). Usually, under these conditions, the size of the cancer is reduced because the removal or inhibition of Tregs allows the anti-cancer T cell immune response to proceed, as is the case with melanoma (Kong et al. 2009).

This evidence suggests that there may be dual expression of the Treg phenotype and the anti-cancer T cell phenotype during the T cell response to GBM. Conversely, Tregs may be in the middle of being converted into anti-cancer T cell responders (Radhakrishnan et al. 2008a) when the Treg phenotype cells are being removed. In addition, conditions within the GBM cancer mass may resemble an inflammatory response, which may cause the conversion of anti-cancer T cells also to Tregs (Curiel 2007). This is why CTL activity can be detected despite the almost complete failure of the T cell immune system to remove the GBM cancer cells.

In contrast, the Tregs that find their way into a melanoma lymphocytic infiltrate are probably of the natural type. Not enough is known about either type of Treg cell, but it is possible that natural Tregs allow for anti-cancer T cell responses and some autoimmune reactions to occur. One possible reason for this is that mature Tregs can be recruited by DCs to become autoimmune effector cells lacking the inhibitory abilities that mature Tregs possess (Radhakrishnan et al. 2008a). For instance, when some melanoma patients respond to their cancers, autoimmune reactions are apparent. During the T cell immune response to melanoma, some of the T cells attack normal melanocytes because of the similarity between melanocyte-specific proteins and the MAAs. The result is vitiligo or loss of pigmentation of the skin (Rosenberg and White 1996). Other cells closely related to melanocytes, such as cells of the nervous system, are also attacked, causing a painful condition called peripheral neuropathy (Visovsky 2005). During these episodes the immune system's attention is focused not only on these two responses, but also on how to stop them.

Both GBM and melanoma cancers elicit a host T cell immune response, which is visible and can be analyzed in the body as well as in the laboratory. Primary melanoma cancers, but not primary GBM cancers, can be removed by the T cell infiltrate. This, despite the fact that both GBM and melanoma cells share T cell-recognizable protein targets, comments on the complexity of the GBM T cell response (Basta et al. 1988). One explanation might have to do with the type of Treg cell that is present in the respective cancer sites. In addition, the recruitment of "adaptive" Tregs from the anti-cancer T cell populations (Liu X et al. 2009a) can contribute to the ineffectiveness of the GBM T cell response. Interestingly, it has been determined that mature Tregs can be recruited to become true effector cells lacking the inhibitory abilities that mature Tregs possess (Radhakrishnan et al. 2008b). This means that the conversion can occur in both directions depending on the immunological conditions at the cancer site.

Another explanation might have to do with the blood-brain barrier, which in some respects has control over the type and number of T cells that enter the brain to attack the cancer. Viral studies focusing on the infection of the central nervous system have revealed that a CD-4 T cell INF-g-dependent reaction may increase the permeability of the BBB (Anirban et al. 2007). This reaction allows for antibodies and T cells to cross the barrier and interact with the cancer. Inhibition of these CD4 cells by either Tregs or antagonistic T cells would restrict entrance and egress of anti-cancer T cells and perhaps "natural" Tregs (as with melanoma), but not "adaptive" tumor-specific Tregs to the brain and thus to the cancer mass itself. "Natural" Tregs are found in melanoma lesions despite the presence of a productive anti-melanoma T cell response. Whatever be the mechanism involved in the failure of the anti-GBM T cell response, these factors must be taken into account during the design of T cell immune-based immunotherapeutic treatments.

Why GBM and Melanoma Are Likely to Respond to Immunotherapeutic Treatments

There are many reasons why both GBM and melanoma are likely to respond to immunotherapeutic treatments. One of the most obvious is that they both express T cell-recognizable proteins that make it easy to design novel treatments. In addition, they both induce regulatory T cell-controlled immune responses (Curtin et al. 2008, Curiel 2007). The ability to manipulate the regulation of the responses is already the focus of immunotherapeutic treatment design. Another similarity between these two cancers is that they develop in stages. The development of a primary tumor stage and the possibility that it will develop into malignant metastatic stage gives scientists time to design immunotherapeutic treatments tailored to that patient's specific tumor or genetic type. Tumor antigens that are specific for the patient's cancer (GBM or melanoma) can be identified and used as targets during immunotherapy.

However, most of these antigens are recognized by CD8 T cells, which at first glance seems to be an advantage. CD8 T cells see the protein on the cancer cell; then they kill the cell. Unfortunately, this is a one-dimensional view of how the T cell immune system works. For years it has been known that both melanoma (Guerry et al. 1984) and GBM (Basta et al. 1988) express HLA-DR in the primary stages of their development. Expression of these antigens allows the T cell immune system (CD4 T cells) to respond to tumor antigens presented by the cancer cell (Alexander et al. 1989). This early CD4 T cell response results in the secretion of lymphokines (Guerry et al. 1987) that induce tumor-specific killer (CD8) T cells to attack the cancer. The response can dictate an agonistic (positive) type of T cell immune response to the tumor (Lau et al. 2005) or an antagonistic (negative) one (Ruppert et al. 1993). This directive given by the naive CD4 T cell is dependent not only on the type of stimulus received by the cell, but also on conditions at the tumor site. It has been determined that within these heterogeneous glioblastomas there are stem cell–like tumor cells, which are radio-resistant and can induce malignant cell growth in non-cancerous tissue. They are called cancer-initiating cells and they elaborate immunosuppressive lymphokines and cytokines that inhibit T cell activation by the generation of Fox3+ Tregs and can induce T cell apoptosis (Wei et al. 2010). This form of control is about predetermination of the anti-cancer T cell response, which in most cases cannot be changed once it starts. Manipulation of these events must somehow be factored into the design of immunotherapeutic treatments.

Current Immunotherapeutic Approaches

Glioblastoma multiforme

Because GBM is usually fatal, there is an intensive effort to develop therapies to increase the cure rate. Clinical trials are underway in Europe to examine the ability of cell-based vaccines to elicit a meaningful T cell immune response to this cancer (De Vleeschouwer et al. 2008). Cell-based vaccines make use of mature macrophages, or DCs. These cells, after their incubation with GBM cell lysates, can process and present GBM proteins on their cell surface so that cancer-specific T cells are generated (Luptrawan et al. 2008). These experiments have been conducted in mice for several years and have demonstrated significant reductions in tumor size and in some cases complete remission. The clinical trials use the patient's own DCs incubated with proteins derived from the patient's surgically removed GBM cancer. Each patient has undergone surgery and chemotherapy. The trials are designed to prevent the inevitable recurrence of GBM. This study demonstrated that the DC vaccine was safe and a significant increase in overall survival was observed. However, stable disease or partial remission was not detected in any of the patients.

Despite the lethal nature of GBM, human immunotherapeutic trials have been somewhat limited as compared to metastatic melanoma. This may have to do with the location of the cancer (brain). Lack of visibility of the cancer during treatment has been a problem in the past, but the visibility of the cancer mass has improved due to new imaging technologies (Lazovic et al. 2008). Another reason for the slow development of these therapies is the short period of time between diagnosis and the death of the patient, which can be from 6 months to 1.5 years. However, the research efforts are moving ahead at a rapid pace.

Melanoma

For melanoma, immunotherapies are focused on the adoptive transfer technique pioneered in the mid-1970s by scientists studying a mouse cancer model. Essentially, the cancer cells (leukemia) were grown in the laboratory. They were inactivated so they could not divide. Then they were cultured with normal mouse T cells for a number of days. After the cells were tested for the ability to kill the cancer, they were transferred into normal mice along with a small number of live cancer cells. Under controlled conditions, it was demonstrated that the mice receiving the immune T cells survived much longer than those receiving un-immunized T cells.

Since then, adoptive transfer has evolved. Two major elements have been added to this type of immunotherapy. The first is lymphodepletion, in which the resident peripheral blood lymphocytes of the patient are temporarily removed by a chemical treatment prior to the infusion of immune T cells. This procedure is thought to remove any immunosuppressive T cells (Tregs and antagonistic T cells) that would inhibit the infused anti-cancer T cells from responding (Gattinoni et al. 2005). The second element is the use of an anti-cancer vaccine during the treatment, which induces anti-cancer T cell proliferation. The lymphokine IL-2 is also infused into the patients to support the maturation and proliferation of CD8 T cells.

In the late 1980s, metastatic melanoma patients were treated with tumor-infiltrating lymphocytes (TIL) isolated from their surgically removed cancer and the lymphokine IL-2. There was a 60 percent increase in the regression rate (Rosenberg et al. 1988). Even though there were problems with the toxicity of the IL-2, these were the first indications that this very lethal form of melanoma (metastatic melanoma) would respond to immunotherapy. Researchers were then able to genetically engineer the TIL to secrete IL-2 (Heemskerk et al. 2008). The trials were performed as above, except that in most cases IL-2 was not administered. Despite this innovation, there was a low response rate. The researchers determined that the TILs had a shortened life span because they spent a longer time in the laboratory to facilitate the genetic engineering phase.

These same researchers went on to the next step, which was to introduce the lymphodepletion technique in the TIL/IL-2 immunotherapeutic regimen (Dudley et al. 2008). This had the effect of increasing the objective regression rate several percentage points above what was observed without the removal of the patient's lymphocytes. The objective regression rate was increased from 51 percent to 77 percent when the patient's lymphocytes were removed by the chemotherapy and total body irradiation. The radiation treatment has the ability to remove bone marrow cells that can regenerate the T cell population, which contain Tregs. Total body irradiation may enhance the regression rate by removing bone marrow cells, which are the source of Tregs. This eliminates the possibility that any newly generated Tregs will inhibit the infused anti-melanoma TIL.

Because the researchers thought the isolation of TILs was laborious and time-consuming, they focused on immunizing the T cells isolated from the peripheral blood of metastatic melanoma patients (Powell et al. 2006). Once those T cells were purified, they were immunized in the laboratory against a melanoma-specific peptide expressed by all metastatic melanoma cells. The patients were lymphodepleted by chemotherapy only. Then they were infused with the immune T cells. A cancer vaccine (same peptide) and high-dose IL-2 were administered. Two patients responded with cancer-induced autoimmune reactions, but none of the nine patients in the study achieved an increase in the objective clinical response. The researchers determined that two critical facts may have contributed to this lack of response.

First, the immunizing protein was recombinant protein made in the laboratory. It has been shown that sometimes these proteins do not generate T cells with high-affinity receptors for the natural cancer protein (Le Gal et al. 2005). Second, the percentage of Tregs in the peripheral blood of metastatic melanoma patients is elevated (Jandus et al. 2008). It is possible that Tregs can be created from the killer T cell pool and still retain the Treg phenotype. The researchers concluded that the removal of these cells may augment the response, but in fact what may or may not happen is a further reduction in the effectiveness of these melanoma-specific T cells, as is the case in GBM (Curtin et al. 2008). The most curious finding of this trial was that two patients had experienced cancer-induced autoimmune reactions. The conversion of Tregs seems to be bidirectional and fluid. T cells with dual phenotypes (Treg and non-Treg or killer T cell) are in the middle; at either end are the true Treg and non-Treg or killer T cell. This may indicate that during the adoptive transfer these two patients had a high number of Tregs that were in the conversion process or were completely converted to non-Tregs during the *in vitro* immunization and lost the ability to control the autoimmune reactions that occur (Radhakrishnan et al. 2008a).

Melanoma has been the focus of immunotherapeutic treatment and many therapies have been used. Some of these are the use of melanoma proteins as vaccines (Bercovici et al. 2008) and the use of T cells genetically engineered to recognize melanoma proteins (Morgan et al. 2006). Melanoma is one of the most intensively studied cancers and these studies are models for the development of immunotherapeutic treatments for a wide variety of other cancers. Some therapies used for the treatment of melanoma promise to improve the cure rate of this cancer as well as many other cancers.

Experimental Immunotherapeutic Treatment Approaches

Some of the immunotherapeutic treatments for GBM and melanoma outlined below may, by the time you read this, already be in the final stages of clinical trials. It is quite clear from the previously outlined facts that both GBM and melanoma are very well suited for immunotherapy. These facts (cancer-specific proteins, T cell regulation of anti-cancer T cell responses, and cancer-induced autoimmunity) are only some of the variables that have to be considered when designing immunotherapeutic treatment protocols.

Glioblastoma multiforme

Cancer cells express altered proteins that are recognized by the T cell immune system as foreign, such as the melanoma-associated protein peptides. Glioblastoma multiforme expresses quite a few of these and researchers have identified a growth factor receptor variant, epidermal growth factor receptor III (EGFRvIII), expressed by GBM tumor cells. This mutated receptor is expressed in some breast cancers and has been shown to stimulate T cell responses that are measured by the detection of antibodies to the EGFRvIII protein. In order to prepare for human trials using this protein/peptide as a vaccine, the researchers immunized GBM patients with the EGFRvIII peptide and detected significant levels of antibodies specific for the protein/peptide (Schmittling et al. 2008). Stage II clinical trials are underway using this variant growth factor receptor protein/peptide.

A similar protein/peptide, Wilm's tumor 1 peptide (WT1), is also expressed by GBM and is the target of phase II clinical trials. WT1 is a kidney-associated protein and is expressed by a wide variety of cancers: breast cancer, leukemia, mesothelioma, and ovarian cancer. In a clinical trial, patients were immunized interdermally with a peptide version of WT1 that is known to be a target of CD8 T cells of their specific genetic haplotype. In other words, the peptide was matched to the patient's genetic type. The cancer mass size was monitored by nuclear magnetic resonance and the measurements were use to determine the effectiveness of the WT1 peptide vaccine. The results were somewhat encouraging in that of the 21 patients who had developed standard therapy-resistant GBM, two patients had a partial response and 10 developed stable disease. However, nine patients continued to develop progressive disease and there were no complete responses. The trial seems to be a failure, but in fact all 21 patients would have progressed to a fatal form of GBM quickly. The statistical information of 9.5 percent overall response rate (with complete or partial response) and 57.1 percent disease control rate (with partial response or complete response and stable disease) is encouraging (Izumoto et al. 2008).

Another approach to the treatment of GBM is to stimulate the immune system in such a way as to awaken an anti-GBM T cell immune response. Researchers are using a lymphokine called IL-4 to stimulate macrophages to become DC antigen-presenting cells. In this case, the gene for IL-4 is inserted into the fibroblast isolated from the patient and as a vaccine adjuvant they use the patient's own inactivated GBM cells isolated from surgically removed cancer tissue. The immunizations were inter-dermal in the thigh. This was the first protocol. In the second, researchers used the IL-4 fibroblast, but instead of cancer tissue as an adjuvant, they used

DCs, which had processed and were presenting the patient's GBM tumor antigens.

In the first protocol, only two patients could participate, but both demonstrated clinical responses and temporary reduction in the cancer mass. However, both eventually died from the disease. In the second protocol, IL-12, which is a product of the GBM protein-expressing DC, was readily detected, demonstrating that the DCs were functioning. However, there was no detection of gamma-interferon, which would have been an indication that CD8 T cells had been mobilized to attack the cancer. In line with these findings was the fact that there was no measurable reduction of the cancer mass or even retardation of the progression of the cancer growth rate (Okada et al. 2007).

Melanoma

One of the hallmarks of melanoma is that there are proteins expressed by the cancer that the T cell immune system can see and respond to. Scientists now realize that when people develop cancer, the T cells with the ability to remove the cancer are already present, but another part of the T cell immune system prevents them from doing so (June 2007). This again has to do with Tregs and their ability to stop immune responses perceived as autoimmune responses. The newest immunotherapeutic protocols for the treatment of melanoma are now divided between controlling the T cell regulation of the anti-melanoma T cell response, using DC therapy to present melanoma-associated proteins, and lymphokine therapy.

Anti-melanoma T cell responses are controlled by Tregs in late-stage melanoma patients (5–10 percent survival rate after 5 years). There is a surface protein (receptor) on the surface of Tregs called cytotoxic T lymphocyte antigen 4 (CTLA-4), which when activated by its ligand (B-7) on macrophages induces these Tregs to inhibit T cell responses of what is perceived to be anti-self (autoimmune) (Sutmuller et al. 2001). This sometimes includes anti-melanoma T cell responses. This reaction to anti-melanoma T cells allows the melanoma to grow. The CTLA-4 protein can be prevented from binding B-7 by the use of anti-CTLA-4 antibodies. If that is done in the course of an anti-melanoma T cell response, the response to the cancer is increased dramatically. However, since these regulatory T cells control other autoimmune T cell responses, the treatments often have serious side effects (Wolchok and Saenger 2008).

As for the immunotherapeutic value of using antibodies against the CTLA-4 protein in patients with metastatic melanoma, the side effects many times coincide with the effectiveness of the treatment. Both anti-cancer T cell responses and autoimmune T cell responses are increased after anti-CTLA treatment. The autoimmune responses documented during these treatments are dermatitis, uveitis, and enterocolitis. Enterocolitis, which causes severe diarrhea and dehydration, is the most common autoimmune disorder. When the patients in this study were separated into those who got enterocolitis and those who did not, a striking difference was seen in the overall clinical response rates. The clinical response rates were 36 percent in those who got enterocolitis and 11 percent in those who did not (Beck et al. 2006).

The intestines in the case of enterocolitis and epidermal cells in dermatitis are not related to nor do they share proteins with neural crest cells such as the melanocyte, from which melanoma develops. This indicates that inhibition of Tregs by this method induces a widespread form of autoimmunity (Holmberg et al. 2005). There is evidence of cancer-specific Tregs. These Tregs are induced only to suppress anti-cancer T cell responses in the presence of that particular cancer, but once activated they secrete lymphokines such as IL-10, which can inhibit non–anti-cancer T cell responses (Liu et al. 2009). It is evident from the facts presented above that this area of research is receiving close attention as a means to eradicate cancers that have been difficult to treat. That research will continue to be intensive until specific functional subsets of the Treg population are identified. That would allow for a more targeted design of immunotherapeutic treatment protocols devoid of the autoimmune side effects associated with the inhibition or removal of Tregs.

Some of the new immunotherapies for melanoma make use of DC vaccines in novel ways. These clinical trials use the patient's DCs and these APCs are pre-incubated with either melanoma-associated peptides or melanoma cell lysates from the patient or from an unrelated patient. In some instances, the lymphokine IL-2 is used to amplify the effectiveness of CD8 killer cells. Clinical trials using melanoma antigen–presenting DCs have met with limited success, but the potential of this type of immunotherapy, if perfected, is enormous. Over the past several years, these trials have resulted in clinical response rates of 5–10 percent, far lower than the response rates of 50–70 percent achieved using lymphoablation, TIL, and IL-2 (Dudley et al. 2008).

A Phase I/II clinical trial was performed using DC-based therapy with or without low-dose IL-2 (Hersey et al. 2008). Sixteen patients were given DCs sensitized to autologous melanoma cell lysates and 18 patients were given DCs pulsed with melanoma-associated peptides representing six well-known MAA. One half of each group was given low-dose IL-2 and the other half was not. The results were two partial responses and three stable disease clinical responses in the nine patients receiving IL-2 and peptide-pulsed DC. There was one partial and one stable in the other half of this group that did not get IL-2. Two of the eight patients developed stable disease in the melanoma cell lysate–pulsed DC group that did not get IL-2 and there were no clinical responses in the corresponding group that did get low-dose IL-2. Statistically, there were no differences between the groups. Since this was a Phase I/II study, all the patients were late-stage metastatic melanoma patients; this type of study focuses mainly on the safety of the treatment.

This type of immunotherapy is constantly being refined because there are many unanswered questions about the variables associated with the therapeutic protocols. Questions abound about the type or maturity level of the DCs, degree or extent of melanoma antigen processing and presentation, and the way the DC vaccine itself is administered. Other questions involve lymphoablation and the inhibition of Tregs by chemical or biological means. Can Tregs be selectively inhibited to avoid damage to normal tissues? There is a need to examine all these variables in a series of clinical trials designed to develop an optimized DC vaccination protocol (Wheeler 2010, Lesterhuis et al. 2008).

Another area of melanoma treatment in which immunotherapeutic clinical trials are being conducted is lymphokine therapy. One of the largest prospective studies, which determined the clinical value of lymphokine treatment in melanoma patients, was compiled over a 21-year period (Smith et al. 2008). The study involved 684 metastatic melanoma patients who were treated with either high-dose IL-2 or high-dose IL-2 and a variety of melanoma vaccines (melanoma-associated peptides). The results revealed that the objective response rate for the IL-2 treatment group was 13 percent and 16 percent for the IL-2/vaccine group. In addition, it was determined that in the IL-2/vaccine group, only patients who had cutaneous but not visceral disease had favorable clinical responses. The study also determined that a particular peptide vaccine, gp100: 209–217(209M), increased the objective response rate versus IL-2 alone from 12.8 percent to 22 percent. All these response rates seem low, but this data was drawn from Phase I/II trials that use late-stage patients.

All of the above-mentioned clinical trials using immunotherapeutic treatments for late-stage metastatic melanoma and GBM have several things in common. First, all of them have been proven relatively safe to administer to cancer patients. This may seem insignificant, but it is a very important aspect of this type of clinical research. The safety of the treatment comes first because without this precaution needless injuries and deaths would occur. Second, the majority of these trials resulted in measurable positive responses. An unfortunate commonality of these trials is the low objective clinical response rates. These rates seem low (13 to 22 percent), but in such trials the safety of the treatment is assessed first, then measurable laboratory responses are detected, and finally the clinical value is assessed on the basis of the objective clinical response.

New Experiments that May Lead to Future Immunotherapeutic Protocols

Several experimental procedures are being performed that might be incorporated into future immunotherapeutic protocols. There are two areas of interest. The first are the DC-based procedures (Wheeler 2010). The second are T cell-based procedures and observations. Some of the procedures use technologies that are relatively new. However, some use basic immunological concepts in a new format. These types of experiments are the basis of new immunotherapeutic treatment design. Most of them are performed either *in vitro* or in rodent tumor models. The intention of these experiments is to point the way toward experimental avenues yet unexplored by other tumor immunologists.

Of the DC-based experiments, the procedure outlined below shows great promise as a future immunotherapeutic protocol. It uses a somewhat recent procedure called DNA vaccination to deliver to the DC a two-part immune stimulating package. There is a receptor (CD205) (Bonifaz et al. 2004) on the DC that when activated by its ligand (in this case anti-CD205 antibody) dramatically increases the endocytic function of these cells. In other words, the activated DCs engulf and process antigens at a much higher rate. DNA is created that transcribes for a fusion protein that is half antigenic peptide (HIV gp41) and half a single immunoglobulin chain (light and heavy chain) of the anti-CD205 antibody. The DNA was injected into mouse leg muscle followed by an electroporation step. Results indicate a more than 10-fold increase in the response to HIV gp41 measured by anti-HIV antibody titer and gamma-interferon production by CD4 and CD8 T cells. This technique shows promise because it directs the vaccine to the cells that are activated to present it. In this case, both CD4 and CD8 T cells were induced to respond, which is a favorable result for an immunotherapeutic protocol (Bozzacco et al. 2007).

Using DCs that were loaded with glioma tumor lysates and a toll-like receptor (TLR) agonist, researchers induced a T cell response to a mouse glioma tumor. The TLRs activate DCs to present antigens upon activation with a TLR agonist such as CpG oligonucleotide. This TLR agonist and glioma cell lysates were encapsulated in a polymer bead. The beads were injected into mice with gliomas or before tumor challenge. It was found that both macrophages and DCs were activated. This resulted in brisk T cell proliferation, but only a slight reduction of tumor growth. However, it was found that Tregs were generated as a part of this process, but when suppressed by anti-CD25 antibody treatment, the remaining anti-tumor T cells were able to completely protect mice from an inoculation of live glioma tumor cells. These experiments, like the previous ones, exploit the DC's ability to become immunologically active, inducing processing and presentation of tumor antigens. In this case, the Tregs probably secreted IL-10 and TGF-B1, which are capable of directly suppressing anti-tumor T cell responses (Grauer et al. 2008a).

With respect to T cell-based experiments that show promise as future immunotherapeutic protocols there are several interesting developments. One development emerges from the fact that a system inherent in our biology triggers cells to die. This is the Fas system and it has two parts. The first is a receptor (FasR) that is expressed by many cells in the body and can be induced in many more by various means. Once a Fas ligand (FasL) makes contact with the receptor, a series of events occur that causes the DNA of the cell to fragment and die. It was determined that GBM tumors do not express FasR, but can be induced to do so by an agent called topotecan (Ackermann et al. 2007). In addition, it has been known that CD3+ T cells express the FasL and can induce Fas-mediated cell death in cells that express the FasR.

Researchers used glioma cell lines and GBM tumor cells isolated from patients and treated them with topotecan and found that they expressed the FasR. Incubation of soluble FasL (sFasL) with the treated cells resulted in significant cell death. More importantly, when CD3+ T cells were incubated with treated cells, increased cytotoxic T cell killing could be detected along with the secretion of gamma-interferon, which is associated with T cell activation/proliferation and cytotoxicity. This technique can be used to induce an anti-GBM tumor T cell response using FasR-expressing tumor cells during an integrated vaccine immunotherapeutic protocol. However, more studies must be done because tumor cells that have been killed by this method, because of the inflammatory nature of that type of cell death, may induce an immunosuppressive response.

Another approach has to do with the source that T cells used to attack the tumor. Scientists have been using umbilical cord blood to isolate T cells to determine how effective these cells can be at killing human neuroblastomas and glioblastoma cells. Mononuclear cells were isolated from umbilical cord blood and divided in half. One half was incubated with IL-2 for the generation of lymphokine-activated killer cells. The other half was incubated with DCs pulsed with tumor lysates to generate a cytotoxic T lymphocyte (CTL) response. Lymphokine-activated killer cells generated from cord blood demonstrated significant cytotoxicity against all neural crest–derived tumor cell lines tested (three lines). The CTLs generated from the cord blood cells demonstrated specific cytotoxicity to the tumor the lysate was derived from. *In vivo*, using a SCID mouse model and a human glioma cell line, CTL treatment of these mice resulted in a specific (as above) and significant increase in survival of mice challenged with live tumor cells. These experiments demonstrate the value of umbilical cord blood as a source of anti-tumor T cells that potentially can be used in lieu of the patient's lymphocytes that have endured chemotherapeutic treatments (Joshi et al. 2007).

In conclusion, cancers of the neural crest have in the past proven to be very difficult malignant tumors to treat. Because of advances in chemotherapeutic treatment, there has been some progress toward improved clinical outcomes. However, this progress has not reversed or even stabilized the high mortality rates of the most malignant of these tumors. Part of the problem with these tumors is systemic and may stem from the fact that neural tissue is a highly "untouchable" tissue as far as the immune system is concerned. Without much effort one could assume that all T cells reactive to neural tissue are rapidly eliminated in the thymus. In addition, there certainly exist controls in the form of T regulatory cells that quickly thwart any neural tissue attack by the adaptive T cell immune system. Looking at this from another point of view could explain why cancers derived from these tissues engender distinct T cell immune responses to the newly developed tumor-associated antigens. The immune system, having never been allowed to see or respond to neural tissue, is awakened by the appearance of altered proteins (neoantigens) produced by the emerging cancerous neural tissue. However these proteins mutate, they now can be called "tumor antigens" and are derivatives of proteins expressed by their normal counterparts. For this reason, the T cell responses are under control by not only Tregs (Grauer et al. 2007, Jandus et al. 2008), but also immunosuppressive T cell responses (Lau et al. 2005, Ruppert et al. 1993) and alternatively activated macrophages (Tiemessen et al. 2007).

The control of the T cell response to these tumors by Tregs merits special attention because it is essential to be able to manipulate this part of the T cell immune system in order to allow the anti-tumor T cell response access to all the cancer cells. As a T cell response is developed against these tumors, there are T cells within this population that attack and kill normal cells that are genetically related to the tumor (Rosenberg et al. 1996). This is cancer-related autoimmunity. Such responses to tumor-related normal cells occur despite the presence of Tregs. Tregs are present in the blood and tumors of melanoma patients (Jandus et al. 2008), many of whom develop these types of autoimmune reactions. In contrast, the removal of Tregs during an anti-tumor T cell immune response can augment the response, but it often results in the development of a number of autoimmune reactions (Wolchok and Saenger 2008). The tissues that are attacked are not related to the cancer in any way (Beck et al. 2006). These are called cancer non-related autoimmune responses. In addition, removal of traditional Tregs by treatment with anti-CD25 antibodies results in autoimmune reactions also related to the target tumor cells. One other critical aspect of this subject is that among cancer patients who display cancer-related autoimmunity with or without treatment, there is a significant increase in survival rate. The question is why? The answer may shed light on the very nature of the anti-tumor T cell response.

Again, malignant melanoma can serve as a model from which to learn what this phenomenon has to tell us about the anti-tumor T cell response. To understand what cancer-related autoimmunity has to do with an improved clinical prognosis, we must examine the normal tissue targets. There are two targets of anti-melanoma T cells during the anti-melanoma T cell response. The first is the melanocyte, and its destruction results in de-pigmentation and is called vitiligo. Vitiligo occurs in some cases independent of an anti-melanoma T cell response. An examination of these situations in terms of an anti-melanoma T cell response may give clues to the relationship between autoimmunity and improved cancer prognosis. The second target is the peripheral nervous system. This painful autoimmune reaction is called peripheral neuropathy. Peripheral neuropathy, too, can be caused by conditions other than the presence of an anti-melanoma T cell response. Again, it may be helpful to understand these reactions in light of what happens in melanoma.

Three situations arise when Tregs are involved with an anti-tumor T cell response. First, during the course of a T cell response to melanoma, cancer-related autoimmune events occur despite the presence of Tregs in the tumor mass as well as the peripheral blood of the patient. Second, if Tregs either are inhibited by anti-CD25 antibodies or are not present during a tumor recurrence where memory T cells respond, the autoimmunity that occurs is of the cancer-related type. Third, when Tregs are inhibited by antibodies against the CTLA-4 receptor protein, the resulting autoimmune reactions are not related to the cancer.

All these scenarios point to the inevitable conclusion that there are unknown subtypes of regulatory T cells. Some of these subtypes have been identified. There is a type of Treg that exerts control over macrophages, controls T cell responses within the tumor microenvironment, and controls systemic responses. However, it is not clear how Tregs that are present in a tumor mass and in the blood of the patient allow tissues related to the tumor to be destroyed while allowing a productive T cell immune response to the tumor. The phenomenon of bidirectional Treg/CTL conversion may explain some of this. The nature of the tumor microenvironment is the key to this pivotal step. CD4 T cells that initially begin to respond to the appearance of a tumor begin to do so under varying conditions.

In an ideal situation (possibly with melanoma), the CD4 T cells meet tumor cells and macrophages, both expressing immunogenic tumor antigens recognized by the T cells. A specific CD4 T cell response ensues to those antigens and the tumor cell is attacked. A natural occurrence in these situations is the recruitment of Tregs that guard against T cell immune responses directed at any other tissues. The need for more anti-tumor T cells increases because of the rapid replication of the tumor, and the consequent induction of CTLA-4 positive T cells actually causes ligation of the B-7 molecule on DCs. The DCs begin to "reprogram" Tregs to become anti-tumor T cells. The loss of the Treg cells results in the loss of control of autoimmune reactions, but only those related to the tumor (cancer-related).

In the less ideal situation (possibly with GBM), CD4 cells arrive at the tumor site and for some time the only immune cells interacting with the tumor have been macrophages. Under these conditions, macrophages that have been exposed to the tumor for extended periods of time begin to differentiate into "alternative activated" macrophages. These cells secrete pro-inflammatory lymphokines and chemokines that induce an environment that can turn CD4 T cells that would become responders into T cell immune suppressing Tregs. Once this occurs, any attempt by T cells to attack the tumor is blocked by several known mechanisms used by Tregs to curtail such activity. Inflammatory conditions have been shown to call Tregs to those sites. The Tregs can immediately suppress inflammatory conditions caused by bacterial infections. Massive inflammation leads to indiscriminate autoimmune reactions. Tregs and other regulatory T cell situations, such as the CTLA-4 situation and antagonistic T cell activation, prevent this rampant destruction of normal tissue (cancer non-related).

Detailed knowledge concerning the complexity of the T cell immune system is needed to understand the anti-cancer T cell responses to GBM and melanoma. From the data presented, melanoma certainly has the most likelihood of first being positively cured by future developments in immunotherapeutic therapy. However, GBM will soon follow it. These two tumors of the neural crest represent some of the most difficult types of cancer to treat. However, there are intensive efforts to develop treatment methods that will improve the cure rate. The treatments outlined here have been derived from knowledge of the inner workings of the T cell immune system. Many of the concepts relating to T cell responses to cancer are derived from basic T cell immunity. Exactly how a T cell perceives the foreign peptide antigen is a simple process with an underlying complexity unknown to scientists several decades ago. Yet, understanding this complexity has yielded tools to allow them to develop treatments and cure some cancers completely. This effort must continue: even though the cancer rate has dropped because of better diagnosis and changes in lifestyles, the mortality rate is still high because of cancers like GBM and metastatic melanoma.

Immunotherapeutic approaches to these cancers are very important because they are so virulent that dangerously high (or toxic) amounts of chemotherapy must be used just to retard their growth, let alone cure them. If done correctly, manipulation of the immune system is a painless alternative. Even though there are side effects with immunotherapy, such as lymphokine toxicity and autoimmune reactions, the effects are manageable. At the same time, side effects such as graft vs. host disease and some autoimmune reactions are treated chemically. Immunotherapeutic treatments for these cancers have been conducted in one form or another for a little over 20 years. However, if the cure rate of these treatments is examined, it appears that there has been little impact on reversing the number of deaths. The encouraging aspect of these treatments is that they are relatively safe and result in positive clinical response rates.

The immunotherapeutic treatments outlined above for both GBM and melanoma are just beginning to be developed. The clinical trials involve late-stage patients and are primarily concerned with safety of the procedure rather than a definable objective clinical outcome. Once these safety issues are achieved, clinicians can move the trials to patients with early disease and combine some of the protocols to significantly increase the clinical response rate to the tumor. Perhaps in the future chemotherapy will be a small component of an integrated immunotherapeutic approach, which includes multiple manipulations of the T cell immune system to achieve a long-lasting cure.

These therapies are at the point where the treatment of HIV was just prior to the development of the current combined therapy, which allows HIV patients to live with the infection instead of dying within five years. The main reason for this apparent difference in time has to do with the fact that cancer is derived from our own cells and is not actually an outside entity such as a virus. There are special conditions that make it difficult to develop a successful treatment for cancer. A virus, because it is not related to self, can be treated by a wide variety of chemotherapeutic agents that attack its unique properties. Many of these properties are not found in our systems, so the chemical treatments are not a lethal threat. However, anti-cancer chemotherapy is toxic not only to the cancer cells but also to normal cells.

This is why the advent of immunotherapeutic treatments for cancers of the neural crest is so important. In the future, the hope is that many of these treatments will be combined in such a way as to completely remove the cancer. The combination of immunotherapeutic protocols must include all stages of the T cell immune system, from the initial CD4 T cell response to the generation of memory effector CD8 T cells. Cancer treatments may be free of toxic side effects, need no surgical procedures, take less time, and leave a permanent memory of the cancer. When that day comes, fear and anxiety about getting cancer will be just a memory.

Chapter 15

The Immunological Hole: Will We Fall In?

A hole, according to Webster's Dictionary, is an opening through something. The immunological hole refers to what our T cell immune system cannot see or respond to. There are two parts to our T cell immune system. The first is the "innate" T cell immune system, the system we are born with. The T cells in our bodies carry genetic information that allows us to respond to pathogens (bacteria and viruses) and cancers without prior exposure. The second part is the "adaptive" T cell immune system. It learns from our exposure to these same threats and develops the actual immunity in the form of memory T cells. Memory T cells are vital to our survival. These two systems save us from repetitive sickness and in some cases death. The problem with the "hole" is that it resides in both systems and in some instances can lead to serious illness or even death. Most bacteria and viruses that we are exposed to will be known to our innate T cell immune system and, if not, quickly responded to by the adaptive T cell immune system, but there are some that cannot be seen by either system.

One obvious example is meningitis. Bacteria and some viruses cause meningitis. The infection causes inflammation associated with the cerebral spinal fluid. In the beginning it causes nausea, vomiting, and severe headaches. However, if left untreated the results can be catastrophic. Death can occur within 72 hours. People who get meningitis are usually, but not always, young. Usually, high school or college students are affected. The T cell immune system of these individuals is unable to respond to the bacteria or viruses in time to prevent the onset of the disease (Ballard 2009).

This appears to be a weakness in our immune system. However, it may be much more than a weakness. It could be either a genetic flaw or a natural phenomenon. In this chapter we examine both possibilities. Initially we focus on how both the innate and adaptive T cell immune systems respond to pathogens (bacteria and viruses) and cancer and we describe the cell types involved in the response as well as the immunological hole.

We examine this response in terms of how our innate and adaptive T cell immune systems work together to successfully combat infections and cancer, then leave behind an immunological memory. We point out why both T cell systems fail to protect some people. Finally, we offer a possible answer to the question: Is the immunological hole real and if so what does this have to do with our survival as a species?

How the T cell Immune System Responds to Infections and Cancer, and the Immunological Hole

Before starting this section we will review the cells of the immune system that participate in the response. The first cell that is pivotal to the response is the macrophage. It is called an antigen-presenting cell (APC) because it actively scavenges the body and has the ability to reduce bacteria, viruses, and cancer cells into protein components called peptides. These peptides trigger the immune system to react. The peptides have to be expressed on the surface of the APC linked to a self-protein called HLA-DR (peptide-self complex). The T cell, in this case the CD-4 T cell, detects this peptide-self complex. It responds to the complex by becoming activated. The activation precedes cell division but includes the secretion of bioactive proteins called lymphokines. One of these lymphokines is called IL-2 and it is needed by CD8 T cells to kill cells expressing a similar peptide-self complex, except that this peptide is attached to the self protein HLA-ABC. The CD-8 T cell kills virus-infected cells before they release other viruses. It also kills cancer cells. Both these target cells must express the peptide-self complex, which is seen as non-self by the CD8 T cell. Finally, the B cell makes antibodies in response to other bioactive proteins secreted by the activated CD4 T cell. The antibodies can attach themselves to bacterial and cancer cells. The macrophage can then ingest these cells and start the process of antigen presentation. Our discussion, for the most part, will center on the T cell immune system because it is largely responsible for the entire immune response. However, the B cell immune system plays an important role by producing antibodies that can attack bacteria, viruses, and cancer.

The bacteria and viruses that cause meningitis are not the only pathogens the T cell immune system cannot see. Viruses cause many problems because of their complexity, or in some instances the lack thereof. The common cold virus is one of them. Scientists have been trying to cure that one for some time. The flu virus is another. It is insidious because it invades our bodies using human DNA obtained from prior exposures to people. As far as the flu virus is concerned, it found a very large hole in our T cell immune system that had to do with regulatory T cells that control autoimmune reactions.

For instance, the flu pandemic in the 1900s resulted in the deaths of millions of people. Australian scientists Peter Doherty and Rolf Zingernagel discovered the HLA proteins. They described how killer CD-8 T cells attacked and killed flu virus-infected cells. They also determined that the millions of deaths during the pandemic were not actually caused by the virus itself, but by T cells attacking normal cells. The virus induced such a strong immune response that the T cells lost the ability to distinguish between self and non-self. Therefore, influenza-activated T cells were attacking normal tissues, which resulted in many deaths (Doherty et al. 2008). This type of autoimmune response is supposed to be stopped by regulatory T cells called Tregs (Longhi et al. 2008). Apparently Tregs were either absent or defective in the 20 million people who died during this pandemic. Millions of other people got the same flu virus, yet survived with the usual flu symptoms. The survivors did not have this hole or absence of a critical T cell population. It is feared that one day such an epidemic will happen again. Immunologists have discovered that Tregs do indeed play a role in the development of the holes that develop in the T cell immune system. They found that Tregs were responsible for keeping the holes from forming during periods of rapid T cell division, as happens during a flu viral infection. Without the Tregs, the T cell immune system is vulnerable (Winstead et al. 2010).

One characteristic of the immunological hole that relates to the innate T cell immune system is that there are gaps in our T cell clonal library (Wolfl et al. 2008). A protein on the surface of the T cell called a T cell receptor (TCR) is designed to attach itself to the non-self peptide (bacteria, virus, and cancer)-self complexes mentioned above. The target cell is usually killed. The T cell population is made up of hundreds of T cell clones. Each one is specific for one or more foreign proteins. The genes for these specific T cell clones are inherited and make up the core of the innate immune system. The problem is that these potential T cell responders cannot respond to all the pathogens known to infect us. However, many people do have them and actually respond without knowing it. The pathogens that infect individuals with this hole in the T cell immune system are everywhere in our environment. Most of the people who have close contact with the infected (meningitis-causing bacteria, for example) usually do not get infected but are immunized as a precaution. Meningitis is usually not passed between individuals for the reasons mentioned above. This hole and its specific relation to bacterial meningitis has been known for some time and, in some instances, can be induced by exposure to some bacterial products. Proteins from the *Staphylococcus aureus* bacterium have been shown to create this hole by eliminating antibody-producing B cells (Silverman et al. 2000). From this evidence it is apparent that there is a hole in the innate T cell immune system because, without these antibodies, macrophages would not be able to process antigen.

What about the adaptive T cell immune system? The adaptive T cell immune system is a somewhat independent entity. However, it is affected by the presence of this hole in the innate T cell immune system because the two T cell systems actually rely on each other. When the innate T cell immune system fails, the adaptive system has a problem. Some of the problems are related to the overwhelming nature of the infectious pathogens or cancer. The onset of some of these events can paralyze the adaptive T cell immune system. This system cannot keep up with an out-of-control pathogen or cancer. These agents can sometimes shut down the adaptive T cell immune system in ways that cause it to inhibit itself (Edwards et al. 2010, Lau et al. 2005). Let us see how both systems work together to fight life-threatening infections and cancer.

How Innate and Adaptive T cell Immune Systems Together Combat Infections and Cancer and Leave behind a T cell Immunological Memory

Usually the T cells of the innate system detect an infection or a cancer. Once this happens, signals are sent to the adaptive branch and the immune system goes into action. Initially, CD4 T cells release bioactive proteins that awaken the entire immune system. Killer T cells are mobilized to attack virus-infected cells or cancer. More bioactive proteins are released that enable many immune cells to participate in the removal of this threat. The most important aspect of this coordinated response is the generation of memory T cells. Without these cells, the T cell immune response would be useless.

The nature of this memory is embedded in the DNA of a small subset of both CD4 and CD8 T cells. The memory T cells actually de-differentiate. That means that they revert to an immature state so that they can exist in the body for years. The signals that allow these few T cells to remain immature are not understood. But it is certain that the development of the memory T cells is what helps us survive. They remain in our bodies for years or even a lifetime. These cells are untouched by aging or other illnesses. A biological reason for this will be discussed shortly.

There are two competing theories about the development of memory T cells. The first is the "linear" theory. As naïve (new) T cells become activated by antigen, part of the process includes the immediate creation of memory T cells after the generation of effector T cells and effector memory T cells. The second is the "parallel" theory, where the result of naïve T cell activation is that memory T cells are created simultaneously with the two other T cell types mentioned above (Holmes et al. 2005). Either way, this change in T cell maturation allows us to respond to recurring cancers as well as to a plethora of bacteria and viruses that we have a natural immunity to or have been immunized against using vaccine treatment.

One of the most exciting aspects of memory T cells is that they have very long telomeres. A telomere is a region of repetitive DNA at the end of a chromosome which protects it from destruction. The length of the telomere indicates how long the cell will survive. The very long telomeres in memory T cells allow them to age much more slowly than most cells in the body (Rosenberg et al. 2009). How this happens during the development of memory T cells is still under scientific debate, but the fact has been proven.

Why Do Innate and Adaptive T cell Immune Systems Protect Some People but Not Others?

Our T cell immune system is supposed to protect us from bacterial and viral infections as well as cancers, but why does it not work all the time? The reason is that there is a genetic disparity between the knowledge contained in the genes that constitute the immune system and the outside world. Infections and cancers intentionally fool us. However, with the proper type of research we can outsmart them. How we deal with this problem can mean the difference between life and death. As mammals, we are created to be diverse. There are many of us. We look different and come from different places. Genetically speaking there are just a few genes that separate the races. A few more genes separate us from primates such as chimpanzees and apes. We are given a set of genes (T cell receptor genes) to protect us from infections and cancer. It too is diverse. If we all had the same genetic make-up, we would all be susceptible to the same diseases. Genetic diversity is closely related to survival (Vandewoestijne et al. 2008). The human species is designed to survive no matter what.

Is the Immunological Hole Real? What Does It Have to Do with Our Survival as a Species?

As humans, we sometimes take our immune system for granted. The system, particularly the T cell immune system, is as much a part of our existence as the oxygen intake into our bodies. We fail to look at the big picture as far as our immune system is concerned. When we see viruses and cancer, we are threatened by the possibility that they are infectious and can be contracted by the simplest means. Of course, there are some pathogens that most of us are susceptible to. Surprisingly, the one we are almost always susceptible to is the common cold virus. It is simple and so mutable that it is one of the most infectious agents. That is, until we become immune after three days of coughing and sneezing. Just think of what would happen if the other serious viruses and cancers operated at this level. We would be threatened as a race, or would we?

In a sense, the immunological hole is a safeguard, or relief valve, against anything like this happening to a deadly virus or cancer threatening every single one of us. The ones who fall into the hole are the infected or affected. The rest must decide whether to work on a treatment or cure for the problem, or wait until nature introduces changes into that pathogen or cancer that can affect the previously unaffected. We usually err on the safe side and begin to study the problem and sometimes even fill in the hole with therapies or medications. In this way, the conditions of the affected people are improved and the unaffected have the knowledge so that they too can be successfully treated to prevent the problem.

After the bulk of this chapter was written, a scientific article was published on how the hepatitis-C virus (HCV) was able to escape T cell immune destruction by exploiting a hole in the T cell receptor library. The scientists determined that HCV produced a mutated protein that was previously recognized by the T cell immune system but had a single amino acid (AA) substitution that rendered it "invisible" to the innate T cell response because no TCR could recognize it. They went on to demonstrate that this phenomenon had nothing to do with APC presentation of the peptide because it was presented in the same manner as peptides without the AA substitution, which were recognized. The article supported the concepts put forth in this chapter and also eloquently stated how this phenomenon is related to the nature of our immune systems. The authors stressed this type of immunological escape mechanism was an important one to study (Wolfl et al. 2008). If the HCV-infected individual's TCR could be genetically engineered to recognize the invisible peptides, a breakthrough in the treatment of the infection may be possible. This concept may sound somewhat far-fetched, but there is clinical evidence to support it.

HCV is an RNA virus that attacks hepatocytes, which are liver cells. It tends to spread slowly through the liver and cause cirrhosis (scarring of the liver) and eventually liver failure. The interesting thing about HCV is the clinical outcome of the infected individuals. There are two outcomes. The first is chronic infection, as happens with 40–50 percent of the infected. Chronic infection is manifested by the appearance of viral-specific immune suppression; viral-specific T cell responses are immediately suppressed by the immune system. This can also lead to a generalized immune suppression, making the patient susceptible to basic infections because of the secretion of immunosuppressive lymphokine IL-10 and the generation of Tregs (Kaplan et al. 2008). The immune-suppressed state is partially due to mutated peptides that the virus makes. These peptides induce only a partial signal to the T cell, which causes the T cell to produce immunosuppressive bioactive molecules that shut down other T cell responses (Lau et al. 2005). This process occurs after prolonged viral replication and is one of the mechanisms RNA viruses use to defeat the T cell immune system. It is highly probable that these patients possess this immunological hole because of the prolonged viral persistence that induced immune suppression and existence of the virus in the liver. After this point, the liver will soon be destroyed without medical intervention, usually in the form of anti-viral medications and, in extreme cases, liver transplant.

In the second outcome, the individual gets infected, viral replication begins, and a T cell response is initiated, just as in the chronically infected patients. In both patient groups there is measurable viral replication and anti-viral T cell activity, but something remarkable happens to the 30–45 percent who are now called HCV-recovered patients. All of a sudden there is no sign of the virus or heightened T cell activity. The virus itself is undetectable in the blood and the liver. Apparently these individuals had a productive T cell response to the virus (Alter et al. 1997). The rapidity of this type of response is critical for the responding patients: when the virus has more time to replicate, immune-suppressing peptides are more likely to be produced. These patients are unlikely to have the immunological hole in the genetic TCR library because the response is swift and complete. The T cell immune systems were able to counter the mutational changes the virus had to offer and killed all the HCV-infected hepatocytes before the infection got out of control.

The immunological hole is not only real, but also a natural phenomenon. Its movement and location throughout the immune genomes of individuals shape our immune system in ways that allow for our survival. This natural occurrence has been operating for thousands of years. The immune system we have now is the result of this moving void. Without it, we would have been extinguished as a species a long time ago. This phenomenon is one of many undiscovered immunological mechanisms embedded in our DNA that has wide-ranging consequences associated with our survival. In fact, it may have shaped our entire genetic make-up.

The relationship between the immunological hole and evolution is more complicated than this attempt to explain it. However, because of this hole, many have died and taken non-disease-related genetic information with them, which means that information was never passed to the next generation. This genetic pruning, so to speak, affected the genetic make-up of our entire race. We are just now beginning to uncover some of these mechanisms in order to understand why our immune systems work, but the genetic ramifications may never be understood. However, with this understanding comes a great responsibility: to use it for the betterment of society. Ethical questions aside, it is fascinating to understand how all this works automatically, without our even knowing it. Not yet, that is.

Chapter 16

Summary and Concluding Remarks

Why Are Advances in Cancer Immunotherapy Important to You and Me?

Immunotherapy will be the ultimate treatment for many forms of cancer. Possibly 20 to 30 years from now, the treatment of cancer will be so radically different from what it is today that it would be like comparing the motorbikes of the late 1800s to the high-tech automobiles of today. The technology of the modern car is so advanced that the average citizen has no idea how to fix it if something goes wrong. If we can better understand immunity to cancer, we can develop more effective treatments for cancer. This book was written to make the subject less complicated. It is an important subject because none of us can say that he or she can never get cancer. Every one of us has the potential to get cancer. We can lower our odds, but if a cancer forms in our body and our immune system cannot remove it, we have a problem. Cancer has affected most people in this country in one way or another. Either you've had it or have known someone who has. The loss of those people to loved ones and relatives is tremendous. But also, our society and its development depend on each individual. The loss of millions of people to cancer has diminished the advancement of science, arts, politics, and other aspects of society. This is why any information about cancer that is understood by a large number of people is important to all of us.

Summary

Chapter 1: Which cells control the anti-cancer T cell immune response?

The T cell immune system is one of the most complex biological systems known to man. It literally means life or death for us all. The T cell immune system is an irreplaceable part of the human body. Each year, new discoveries are made about how T cells behave and their relationships to the other cells in our bodies. Chapter 1 conveys the interrelationships between the T cell immune system and other immune cells in the backdrop of an anti-cancer T cell response. It demonstrates how immune cells work in concert to allow their enforcers (T cells) to destroy cancerous cells. All the cells (including the cancer itself) play a part in this battle between the T cell immune system and cancer. This conflict shifts back and forth, leading to the death of the cancer or the person. Many times, the initial interaction between T cells and cancer cells is so swift and decisive that almost immediately there is only one of two outcomes: complete destruction of the cancer cells or the disabling of the T cell immune system by the cancer so that the cancer takes over the body. This can happen at such an early stage that it cannot be measured. The T cell immune system works rapidly and automatically but is prone to mistakes and missteps due to the biologically illogical development of cancer. This information is provided to lay the groundwork for the reader to understand the intricacies involved in the development of cancer immunotherapy, where these mistakes are corrected by the manipulation of the forces that empower the T cell immune system to fight cancer.

Chapter 2: Antigen presentation

The process of antigen presentation is an essential function of T cell response to cancer. Almost every aspect of the T cell attack is related to antigen presentation and the cells responsible for it, the monocyte/macrophage series. The ability of cells to present protein/peptide antigens to T cells was thought to exist only with the cells of the monocyte/macrophage series. However, it has been discovered that a wide variety of other cells can perform this function, including the cancer cell itself. Antigen-presenting cell diversity helps the T cell immune system recognize antigens available throughout the body. This allows the T cell immune system to follow the "bread crumbs" to the cancer site. In some cases, the cancer site is the bloodstream. This subject is also addressed in the discussion of leukemia in Chapter 6.

Once we know the mechanistic aspects of this process we can appreciate the elegance and biological intricacies of a system that practically functions on automatic. The various nuances that go into what appears to be a straightforward process of breaking down foreign proteins into peptide subunits and putting them on the cell surface denote a flexibility of the system that allows it to handle previously unknown viruses and cancers. Because this is an essential aspect of the anti-cancer T cell response, new research in this field was reviewed. The application of these new findings in antigen presentation research will have an immediate impact on the development of novel immunotherapeutic treatments for cancer. Treatments developed from such research will allow the T cell response to cancer to attack more specifically. These refinements may lessen the use of traditional chemotherapeutic agents, which sometimes have adverse side effects.

Chapter 3: T cells and cancer cells: Components of a new anti-cancer therapy

Chapter 3 is concerned with the direct interaction between T cells and cancer cells. It goes into some detail about the origins of both cells and how they are designed to battle each other. The chapter also points out the frailties both cells possess that make this conflict complex and cause problems during the response: for example, the T cell has vulnerabilities that can "accidentally" damage normal tissue during the response to the cancer, and the cancer itself inadvertently signals the T cells to attack it. Two other problems are that T cells regulate themselves during the anti-cancer response and that the cancer cell deceives the T cell by making it respond as if to a normal cell. These activities are examined in the light of the development of new T cell immune-based therapies for cancer.

Chapter 4: Anti-cancer vaccines: New approaches

The concept of a vaccine is simple: once you have been treated with the vaccine you are immune to what the vaccine was made of. But the vaccine story also involves how vaccines were developed, how they have saved millions of lives, and how our society has been shaped by vaccine development. Chapter 4 focuses on vaccine development for the treatment of cancer, but it takes a turn toward T cells that inhibit the very cells that are supposed to attack the cancer. There is a description of what a vaccine is and how the whole concept of preventing a disease got started. The cellular aspects of how modern vaccines elicit a T cell immune response are highlighted so that the complexities of the response to a vaccine can be understood.

Successful use of an anti-cancer vaccine is a proactive process rather than a preventative measure. Unlike the use of a vaccine to prevent papilloma virus infection that causes cervical cancer, the modern anti-cancer vaccine actually induces a T cell response to a growing tumor. This use of anti-cancer vaccines as an immunotherapeutic treatment is complex and is susceptible to control by the T cell immune system itself. A good part of Chapter 4 has to do with that very subject. Regulatory T cells, agonistic/antagonistic T cell responses, and dendritic cell–mediated control of T cell/Treg conversion are all areas in which the vaccine therapy for cancer may fail. Particular attention is paid to T regulatory cells because they are present during strong secondary T cell responses, which vaccines are designed to create. The presence of these cells during such responses can render the anti-cancer vaccine practically useless. Only when these issues of T cell regulation of vaccine-induced anti-cancer T cell responses are dealt with can these vaccine-based approaches be successful. A final discussion on the possibility that anti-cancer vaccines should be directed at T regulatory cells instead of at the cancer brings the issue of anti-cancer vaccines into focus.

Chapter 5: How and why do we get cancer and can immune-based anti-cancer therapy save us?

Will cancer become a routine, manageable illness like some viral infections as we perfect immune-based treatments for it? This was one of the main issues this book was written to address. Can all the combined knowledge, experimental experience, and scientific and medical observations about our T cell immune system be put together to develop a safe, painless, and final treatment for cancer? For the first time in our scientific and medical history, we have developed a critical mass of information that is about to explode into life-saving treatments for cancer. Looking over most of this information, one cannot see a future where contracting cancer would be the same death sentence it was just 10 years ago. To the contrary, some of the most difficult types of cancer to treat will yield to this approach of using the patient's own T cell immune system to bring about the destruction of their cancer. This information about our T cell immune system, and the immune cells that support it, is extensive and complex. There are many activities our T cell immune system is involved with that seemingly have nothing to do with fighting cancer. What scientists have done is harness these activities and manipulate them to make the T cell immune system attack the cancer.

Because the development of immune-based cancer treatment involves many types of cancer research, Chapter 5 includes a brief history of sociological and governmental changes that occurred in the early 1970s. Next, it describes the basic elements that cause cancer and how a cancer cell develops. An important aspect of this type of anti-cancer therapy is how the T cell immune system relates and responds to cancer. Another related question is: How are other immune cells involved in this activity? A critical consideration is how the cancer can undermine or subvert the T cell immune system to grow and spread throughout our bodies. The last two areas are covered in detail because of their importance to the implementation of immune-based anti-cancer treatment. Traditional chemotherapy is examined from a historical point of view and is brought up to date by looking at new chemotherapeutic approaches.

A brief description of just what immunotherapy is and how it was first developed sets the stage for the use of this relatively new anti-cancer therapy in humans. The chapter describes just what immunotherapeutic trials are, how they are performed, and the various levels of treatment. It discusses the meaning of these levels of treatment from a clinical point of view so that one can appreciate the cautious approach taken toward experimentation involving humans. The meanings of the various clinical outcomes of each stage of this process are explained to demonstrate the step-by-step process of modern clinical trials. An important aspect of the trials is addressed: patient participation and informed consent to participate in an experimental procedure.

The three major immunological areas covered by the clinical trials are discussed in detail. In the future, these areas will be among the first to be approved for standard treatment. Each area is examined for the mechanisms involved, the possible outcomes, and their potential benefit to the patient. Specific types of immunotherapeutic clinical trials are mentioned. These are based on the type of immune cell used to facilitate the T cell immune system's removal of the cancer. Genetic alteration of these cells is an important technique used during the deployment of these studies, so considerable attention is paid to how it is used and when. Because of the extensive history and use of chemotherapeutic agents to treat cancer, the use of chemotherapy and immunotherapy together is covered. This part of the chapter gives specific detailed descriptions of how both these therapies used together alter the T cell immune system in such a way as to induce it to attack the cancer. Finally, the future of immunotherapy is discussed from the point of view of individuals and society.

Chapter 6: Immunotherapeutic approaches to the treatment of leukemia

Chapter 6 is dedicated to the efforts to cure leukemia through immunotherapy. Leukemia is in a sense a representation of the human struggle against cancer. It is a very special form of cancer that originates in the bone marrow, which is where our cellular immune system is born. Even more important, some of the immune cells that are supposed to protect us become a part of the leukemia. There are special problems related to the ability of the T cell immune system to fight this cancer. For instance, when a lymphocyte becomes a leukemia cell, the immune system must attack a part of itself. This is a serious problem because the T cell immune system was not designed to respond in such a way and on such a large scale. T cells can attack other T cells (in the case of Tregs), but leukemia has nothing to do with autoimmunity. There is a need to attack the cancer right away, but that does not happen because the T cell immune system has to make adjustments and is triggered by certain events such as the release of lymphokines and proteins that signal the presence of a cancer. These events occur quite late in the development of leukemia because often the early leukemic cells function normally and it becomes a matter of how many there are and not that they are dysfunctional.

This chapter describes the nature of leukemia and the amazing attempt by the T cell immune system to respond to it. The definition of leukemia leads into the problems faced by the T cell immune system. The immunotherapeutic treatments are described as they relate to their role in supporting the chemotherapeutic treatment of leukemia. The next section shows how scientists have been able to redirect the T cell immune system to recognize and attack the leukemia. Finally, the future of these new therapies and the impact they may have on our society are discussed.

Chapter 7: Cancer-induced autoimmunity

Cancer-induced autoimmunity is something that most people (cancer patients included) do not think about. It can be considered a secondary disease affecting those who have cancer. However, its symptoms and prognosis are anything but secondary in terms of the severity and length of the autoimmune episode. Many patients suffer from cancer-related autoimmune reactions before, during, and after the cancer has been resolved by treatment. In fact, many people first learn they have cancer after the investigation of one of these episodes. This is because as T cells attack the cancer, they sometimes attack normal tissues related to the cancer. Chapter 7 highlights one of these autoimmune responses associated with melanoma as a "classic" representation of cancer-related autoimmunity. There is a simple explanation of general autoimmunity, including the mechanism of its development and the immune cells involved in the response. A modest evaluation of the treatments for generalized autoimmune disorders is put forth to point out some potential dangers based on the resulting T cell responses. Also, an aging or weakened immune system lends itself to the development of autoimmunity because of an age-related reduction in the number of T regulatory cells or Tregs. Often, in older patients, cancer-related autoimmunity becomes a major issue.

Tregs are responsible for controlling autoimmune responses. This chapter describes the dichotomy in controlling the inhibitory activities of Tregs during an anti-cancer T cell response. Side effects may leave the patient vulnerable to autoimmune diseases, like lupus, diabetes, and neural damage. There has to be a specific way to identify Tregs responsible for regulating T cell responses for a particular type of tissue. In the case of Tregs inhibiting cancer-specific T cell responses, this identification will allow scientists to separate the Tregs preventing cancer-induced autoimmunity from those interfering with the T cell response to the cancer. Once they have done that, they can remove one group of Tregs (anti-cancer T cell response inhibitors) and stimulate the other (autoimmune Tregs). The system will then work the way it is supposed to.

Chapter 8: Can we solve the cancer stem cell program and cure cancer using T-cell-based immunotherapy?

Chapter 8 highlights a recently discovered fact about cancer cells. For the first time cancer is viewed not as a single type of cell, but as a population of malignant cells born out of a cancer stem cell or cells. Each stem cell has the ability to mutate into distinct subpopulations of cancer cells. These subpopulations have unique characteristics that in many cases are recognized by the T cell immune system. The ability to change is called differentiation and is at the heart of the stem cell nature of the cancer. This function is a problem when it comes to the T cell immune response to cancer. The nature of this stem cell quality is examined by looking at the individual steps a normal cell goes through just before it becomes cancerous. Once treated, the cancer stem cells are changed by the T cell response to them. They are also changed by the accumulation of DNA mutations known to occur in cancer cells. Because of the continuous differentiation of cancer stem cells, they are able to sometimes outmaneuver the T cell immune response so as to become almost "invisible". This chapter outlines the strategies used by both the cancer cell and the T cell immune system in their pitched battle. An experimental approach is proposed to sequentially understand the cellular program that operates within the cancer stem cell population. This experimental model is designed to induce genetic changes in the stem cells by immune T cells in order to discover a predictable pattern of changes in the cancer cell's expression of proteins that potentially can alter the T cell immune response to them. Once that is achieved, the information can be used to design immunotherapeutic treatments.

Chapter 9: Translational immunology and its impact on anti-cancer T cell immunity and the development of immunotherapeutic treatments for cancer

Intercellular pathways are at the root of the functional behavior of all immune cells. The induction of these pathways dictates the direction, intensity, and length of the functional activity of tumor-responding immune cells. These pathways and their activation control immune cell behavior on a very basic level. Understanding and manipulating the pathways within immune cells can lead to vastly improved immunotherapeutic treatments for cancer.

To simplify a complex subject, Chapter 9 describes what happens inside the T cell when the T cell receptor (TCR) encounters an immunogenic peptide associated with the MHC molecule of the APC. It describes how the receptor and associated proteins are configured on the T cell surface and just below the cell membrane. In addition, the role these proteins play during the engagement of the TCR with the peptide complex is examined in detail. There is a simplified description of the biochemical reactions that result in a signal being sent to the nucleus of the cell. These pathways intersect within the cell in a cascading fashion, resulting in specific instructions that the nucleus must carry out. Those instructions may range from the induction of lymphokine secretion to T cell proliferation, T cell suppression, and a number of other actions performed by T cells during an immune response. During this interaction between T cells and APC there are other TCRs and ligands that are engaged by the APC. These engagements lead to the activation of the same pathways or completely different ones. All of that leads to a specific action the T cell must perform.

The intercellular pathways examined in this chapter control the function of two important cells that support the immune response: the DC, which starts, modifies and stops the T cell response to a tumor, and the NK cell, which has a complex relationship with both DCs and T cells during the anti-tumor T cell response. The intercellular changes that occur during DC differentiation are highlighted because without them anti-tumor T cells would fail to attack the tumor. After these morphological changes, a series of transcriptional events result in irreversible changes in the DC's appearance and function. The next aspect of DC anti-tumor immunology, associated with intercellular activation signals, has to do with the interaction of MHC class II protein of the DC and the lymphocyte activation gene-3 protein or LAG-3. This interaction induces inhibition of DC maturation and a reduction in the ability of DCs to activate naïve anti-tumor CD4 T cells. The intercellular events that govern this inhibitory activity of DC involve the coalescence of three separate intercellular membrane domains, which result in the formation of an inhibitory pathway associated with a kinase substrate called ITAM. The initiation of this pathway by the interaction of LAG-3 and MHC class II results in a negative signal being sent to the nucleus of the DC. This signal initiates the reduction of DC cell surface proteins, which facilitate the activation of naïve CD4 T cells.

Intercellular events that occur within the DC control its ability to convert Tregs to autoimmune cytotoxic effector cells. This very important function that DCs perform can lead to the eradication of the tumor or, under special conditions, to autoimmune toxicity, where vital organ tissue systems are attacked by T cells. The intercellular events that lead to both of these activities follow the same pathway. All this begins when a receptor (B7.1 or CD80) on the DC is crossed-linked by either a B7.1-specific antibody or the natural ligand expressed by T cells, called PD-1. Cross-linking activates an intercellular protein called DAP12 that contains a kinase substrate. This activates a kinase called syk, which phosphorylates phospholipase C. In short, this pathway allows DCs to create these autoimmune effectors. Under the right conditions, these effectors can attack the cancer because they were once tumor-specific Tregs that have a TCR for the tumor. However, they can be created under conditions where antibodies are being used to eliminate Tregs. When this treatment is performed, Tregs that were protecting normal cells and have receptors for them begin to attack those cells after conversion to autoimmune effectors. At this point, an inhibitor of the syk kinase can be used on DCs during immunotherapy to prevent them from creating these harmful autoimmune effectors.

Intercellular pathways of the NK cell that are responsible for the mobilization of perforin granules, and the resulting death of the target tumor cell are described. A pathway unique to the NK cell, which has to do with lymphokine secretion, is a powerful weapon these cells use during the anti-tumor T cell response. The NK cells use lymphokine secretion to accelerate the anti-cancer T cell response and to prevent graft vs. host disease (GVHD) during bone marrow transplant therapy. How this works is still being studied, but it is known that if host NK cells begin to make certain lymphokines early in the transplant phase, these lymphokines suppress the ability of the donor's T cells to attack the normal tissues of the host, thereby limiting or attenuating the effects of GVHD. Therefore, controlling the intercellular pathways that govern lymphokine secretion by NK cells is very important.

Finally, recently discovered intercellular pathways in immune cells potentially can have a positive impact on the development of immunotherapeutic treatments for cancer. In a system called "immune surveillance", T cells and NK cells constantly make contact with most cells in our bodies. In some cases, where a T cell may come in contact with a developing cancerous cell, an intercellular oncogene protein (H-Ras) is transferred from the cytoplasm of the suspect cell to the cytoplasm of the T cell. The accumulation of H-Ras, which is called a G-protein, in the cytoplasm of a T cell induces T cell activation. Therefore, if the T cell is in contact with this developing tumor, the cell will transfer a significant amount of H-Ras to the T cell, which then begins the activation cycle. This can be a problem for the T cell: if the cell is a fully formed tumor cell and devoid of MHC proteins, by mutation, it will transfer too much H-Ras to the T cell. Without the T cell recognizing any tumor cell–bound MHC proteins, it will be suppressed. However, there is no such problem with the NK cell because it does not need expression of MHC proteins by the tumor cell to kill it.

Regulation of T cell motility is important for treatment of cancer by immune-based therapies. During immunotherapy using DCs as tumor antigen APC, the patient's DCs are created in the lab and incubated with tumor antigens from their tumor. The tumor antigen–presenting DCs are infused back into the patient with an adjuvant, such as a lymphokine. Once the cells reach the tumor site, they are supposed to activate tumor antigen–specific naïve CD4 and CD8 T cells. However, because Tregs are faster than naïve T cells, they intercept the immunotherapeutic DCs and begin to shut them down. This shutdown of DC maturation and function is contingent on the number of Tregs at the tumor site. If there are very few Tregs, naïve T cells can begin to be activated by the tumor antigen–presenting DCs, but if the number of Tregs is high they not only shut down DC function, but also make it physically difficult for naïve T cells to make contact with DCs. The pathway that is about to be summarized is important because it can increase the odds that naïve T cells can find DCs before Tregs do.

The pathway begins with the contact between the lymphocyte function-associated antigen-1 (LFA-1) ligand on T cells and its receptor the intercellular adhesion molecule-1 (ICAM-1) protein expressed by the DCs. Both naïve T cell and Tregs express LFA-1, but Tregs express more because they are activated T cells. Initially it was thought that the interaction between LFA-1 and ICAM-1 resulted only in DC inactivation. However, when the LFA-1 protein makes contact with the ICAM-1 protein, it generates an intercellular signal within the cytoplasm of the T cell. This signal is generated only when the CD4/CD8 proteins of the T cell make contact with the MHC class I/II proteins of the DCs. A series of intercellular events are described that increase T cell motility dramatically. The idea is to artificially increase the activity of the intercellular activation events in naïve T cells and use these altered T cells during immunotherapy with tumor antigen-presenting DCs. This is just one way translational immunology can be used to augment the effectiveness of treatments. This summary was presented in more detail because some of the chapters use specific examples in which the highlighted immune cell functions through these pathways and its manipulation increases its performance during immunotherapeutic treatment.

Chapter 10: Dendritic cells, immunotherapy and the anti-cancer T cell response

Dendritic cells (DC) play a pivotal role in the T cell response to cancer because of their heightened ability to present tumor antigens both to activate CD4 T cells and to induce anti-tumor cell cytotoxicity in CD8 T cells. However, DCs not only control the induction of a productive anti-cancer T cell response, they also prevent one from occurring. They do this by creating T regulatory cells (Tregs), which under certain circumstances view an anti-cancer T cell response as a response to self. Tregs then go about inhibiting not only anti-self T cell responses (created accidentally by the anti-tumor response), but also T cell responses targeting the tumor cell itself. This activity by DCs is driven by the environment they find themselves in during an active anti-cancer T cell response. That depends on factors that have to do with the ability of the cancer cells to induce an inflammatory response. If these conditions exist when DCs arrive at the tumor site, they will transform T cells in that area into Tregs. Tregs exist to halt autoimmune reactions that are harmful to the body no matter how they are created. Because inflammation is a precursor to autoimmunity, DCs respond by creating Tregs.

Dendritic cells also communicate and, in some instances, control immune cells other than T cells. For instance, if they encounter an anti-tumor T cell response where some of the tumor cells have been killed, they sample the tumor antigens from the dead cells and immediately report back to the nearest draining lymph node. Aside from activating tumor-specific CD4 T cells, the DCs also interact with NK cells. From that point on, NK cells leave the lymph node, go to the tumor site, and kill even more tumor cells. As more tumor cells die, more DCs are activated and again go back to the same lymph node, activate more CD4 cells, and release even more NK cells to the tumor site. This is a continual activity and usually results in the swelling of the draining lymph node because of the build-up of so many lymphocytes involved with fighting the cancer.

This chapter examines how and why DCs control other immune cells, and why other immune cells can control DC functions. These interactions are two-way because DCs induce immune responses and also shut them down when the threat has passed. They secrete probably more kinds of lymphokines and cytokines than any other immune cell in the body. Depending on the circumstances induced by different types of immune threats (for example, viral infection versus cancer), the DCs secrete lymphokines in specific combinations to control several different subsets of T cells and NK cells to perform specific tasks.

This chapter also looks at the extensive expression of receptors and ligands DCs use to communicate with during cell-to-cell contact. These encounters can direct the DC to perform specific functions and also affect the other cell. Sometimes, a two-way signal is generated during contact that causes both cells to behave differently. This altered behavior by both cells has a negative impact on the use of DCs as vehicles of active anti-cancer immunotherapeutic treatment protocols. Many problems arise when cancer patient DCs are used to induce an anti-cancer T cell response. Because DCs are best at presenting immunogenic tumor antigens to T cells, they are thought to be the ideal cell to use for the treatment of cancer. This activity can be improved by inhibiting intercellular pathways that cause DC to fail in this task. The pathways in DCs are activated by Tregs already in the tumor site. The Tregs are drawn to the DC by the lymphokines secreted by activated DCs (created in the lab for anti-cancer immunotherapy treatments) and by surface proteins expressed by both cells. This quick attraction to DCs by Tregs prevents naïve tumor antigen–specific T cells from communicating with the DCs, thereby inhibiting the formation of a productive anti-cancer T cell response.

Chapter 11: Regulatory T cells, dendritic cells and the success of immunotherapeutic treatments of cancer

Chapter 11 describes something that is more than likely responsible for most cases of cancer going out of control and killing the patient. Cancer causes damage, but it happens in the first place because of our own T cells. The T cells responsible for the escape of the cancer cells can be the very same ones that a moment ago were killing cancer cells. A situation in which T cells (regulatory T cells or Tregs) are preventing anti-tumor T cells from attacking the tumor cells can be solved and the T cell immune system may be able to completely remove the tumor from the body. However, that is easier said than done. The T regulatory system is an essential part of our immune systems. Without it we would suffer from autoimmune disorders, some benign and some more deadly than cancer. Yet, it prevents the removal of cancer by the T cell immune system. To develop modern immunotherapeutic treatments, we must understand how this happens and control it.

This chapter deals with the creation, control, and conversion of Tregs in the context of the anti-tumor T cell immune response. It has recently been discovered that Tregs are converted into autoimmune effector T cells. This finding has serious implications in that these effector cells, converted by DC from tumor antigen–specific Tregs, can be a source of anti-tumor T cells. However, the conversion of tissue-specific Tregs into these effectors can induce autoimmune reactions not related to the tumor. For instance, if pancreatic antigen–specific Tregs, which protect pancreatic cells from T cell attack, are inadvertently converted by this process they will immediately attack pancreatic cells and cause diabetes. Cancer patients often suffer from a variety of unrelated autoimmune disorders potentially caused by this newly discovered conversion event.

The chapter examines the conversion of either naïve or antigen-specific T cells into Tregs by DCs. Dendritic cells convert anti-tumor T cells in a tumor mass into Tregs because of inflammatory conditions induced by the presence of the tumor cells. Inflammation is a precursor to autoimmunity and these conditions "trigger" the DCs to secrete specific lymphokines and cytokines that act upon T cells, causing them to take on the phenotype of a Treg. This does not require the T cell to divide. The conversion after exposure to these lymphokines takes time, but ultimately a Treg cell is formed that prevents anti-tumor T cells from attacking the tumor. Also, once a T cell is converted into a Treg it secretes the same lymphokines that the DCs secreted to make it. In other words, the creation of Tregs induces the creation of even more Tregs because of their lymphokine secretion profile. This phenomenon will have serious implications as we discuss how Tregs control DCs. Apart from lymphokine secretion, DCs also convert T cells into Tregs by cell-to-cell contact with T cells.

There is an examination of the methods Tregs use to prevent T cells from responding to the tumor. The lymphokine secretion mentioned above not only creates Tregs, but can also retard T cell proliferation at a distance. Tumor antigen–specific T cells need to proliferate in order to function properly, but Tregs do not because they are already a form of activated T cell. This is one weapon; another is T cell cytotoxicity. Tregs can kill tumor antigen–specific T cells if they come too close to the tumor. The Treg is attached to the tumor by a tumor antigen–specific TCR and can intercept an anti-tumor T cell attack. Tregs have this TCR because they used to be tumor antigen–specific T cells before they got converted into Tregs.

Tregs also control DCs and that affects immunotherapeutic treatment of cancer when DCs are used as cellular vaccines. Because DCs present antigen better than any other APC, they are the perfect cells to use to increase a patient's anti-tumor T cell immune response. However, clinical trials using DCs in this manner have met with little success: when the tumor antigen–presenting immunotherapeutic DCs are infused into the patient and find their way to the tumor site, they come under the influence of resident Tregs. Tregs control DCs through lymphokine secretion and cell-to-cell contact. The cell-to-cell contact involves a ligand on the Treg surface called LFA-1. Its receptor on the DC is called ICAM-1. When the two join, the cells are glued together until the Treg lets go. It will let go only when it detects the reduction of surface receptors (B7.1/CD80 and B7.2/CD86) and the absence of DC-secreted IL-4, which is needed to activate tumor antigen–specific naïve T cells. The Treg effectively blocks the ability of the DCs to activate tumor-responding T cells. Three other methods of cell-to-cell contact are described, including a recent finding that Tregs can kill tumor antigen–presenting DCs exactly the way CD8 CTL kill the tumor. Tregs control DC function through secretion of both IL-10 and TGF-B. The two lymphokines (TGF-B is a cytokine) induce the conversion of T cells into Tregs. When immunotherapeutic DCs arrive at a tumor site, the Tregs bathe the DCs in these lymphokines and the DCs themselves begin to secrete the same lymphokines in response. The very cells (DC) that were supposed to generate anti-tumor T cells now convert non-Treg T cells into Tregs. This is a serious problem for anti-tumor immunotherapy in general, but specifically for therapies using antigen-presenting DCs as cellular vaccines.

The combined effects of these conversions and Treg control of DC makes it difficult for the treatments to work effectively because they can actually accelerate tumor growth. This acceleration of tumor growth may not even be measurable because one part of the tumor is destroyed by the treatment, while the other part is released from control by a diverted immune system. This is just a part of the complex nature of immunotherapy.

Immunotherapy treatments can be improved by first focusing on intercellular activations inside DCs and on ways in which the activation pathways can be altered to make the DCs resistant to control by Tregs. The DC's ability to convert tissue-specific Tregs into harmful autoimmune effector cells can be prevented by treating immunotherapeutic DCs with a kinase inhibitor called syk. This blocks the pathway leading to the production of the anti-self effector cells. A second conversion event involves the DC's ability to convert T cells into Tregs. This can be advantageous in the short term because it would allow DCs to induce a productive anti-tumor T cell response. In the long run it may be harmful to the patient because Tregs control T cells that cause cancer-induced autoimmune reactions, which can be deadly. However, as a temporary approach to immunotherapeutic treatment it may be effective.

There are two situations in which DCs induce the conversion of T cells to Tregs. First, during a viral infection the TCRs of anti-viral T cells begin to mutate and lose the anti-viral specificity, but retain the anti-self element. This is a common occurrence because recognition of virus-infected cells by effector T cell requires both to be present. Tregs are then created (converted) that eliminate the anti-self effector cells. Second, when a tumor forms, another factor is involved and that is the creation of "neoantigens" by the developing tumor. These antigens are slightly altered normal (self) antigens. Nonetheless, they are responded to by T cells and NK cells. Because neoantigens are similar to self antigens, Tregs are created at least 10 times as fast as in the earlier situation. They seem to expand as the pre-malignant cells do. However, there is a way to treat DCs before they are used to fight a cancer that has developed this way. Ordinarily if tumor antigen-presenting DCs used for immunotherapy found their way to a tumor mass, created as above and full of Tregs, the DCs would be recruited by the resident Tregs to produce even more Tregs because of their power over DCs. The DCs have toll-like receptors (TLR). The receptors, once engaged by their ligands, lipopolysaccharides (bacterial cell membrane debris) and unmethylated CpG DNA motifs (free-floating bacterial DNA), cause DC to induce Tregs by lymphokine secretion. That involves the p38 MAPK intercellular pathway. If this pathway is blocked by the use of specific p38 inhibitors, the DC not only are unable to induce Tregs, but also begin to secrete IL-12, which can induce naïve tumor antigen–specific CD4 T cells to become activated. If immunotherapeutic DCs are first induced through the TLR proteins and then treated with p38 inhibitors, they will be resistant to the influence of resident Tregs in the tumor mass and will also be able to induce a productive anti-tumor T cell response by the secretion of IL-12. This is one of the ways immunotherapeutic treatments involving DCs could be augmented because DCs are very sensitive to these Tregs already present in the tumor mass.

Toward the end of the chapter there is a description of two lymphocyte proteins that interact with proteins on the DC, which inhibit their function. These are LAG-3, a ligand whose receptor is the DCs MHC class II protein, and LFA-1, whose receptor is ICAM-1 on the DC. The LAG-3/MHC class II interaction creates a coalescence of cell surface receptors whose trans-membrane proteins create an ITAM inhibitory pathway, which leads to reversal of DC maturation and the inability to induce an anti-tumor T cell response driven by tumor-specific naïve CD4 T cells. The LFA-1 and ICAM-1 proteins are significant in the use of DCs during immunotherapeutic treatment of cancer because of two factors (Onishi et al. 2008). The first is that this interaction between LFA-1 and ICAM-1 is a high-affinity association between the two cells. This means that the Treg and the DC are locked together. Tregs quickly detect the DCs and make contact. All the other protein-to-protein interactions that occur between the two cells follow this step. The Tregs get to the DCs before naïve T cells can and physically crowd out naïve T cells. The LFA-1/ICAM-1 interaction results in the reduction of the ability that DCs have to activate tumor-specific naïve T cells. A possible solution to this problem is to treat the DCs used for therapy with either LFA-1 blocking antibodies or a soluble form of LFA-1 itself.

Tregs interfering with the anti-cancer T cell immune response are a problem during the T cell immune response to tumors in general, and a major problem during immunotherapy for cancer. Because Tregs are complex and absolutely necessary as a part of our immune systems, they must be manipulated with caution during immunotherapy. Tregs and DCs are the main cellular players involved with the inhibition of anti-cancer T cell responses. Because the DCs secrete so many lymphokines, cytokines, and chemokines, they can present problems when used during therapy because of their influence on normal cells.

Cancer immunotherapy is the next "frontier" of cancer treatment and the solution to this Treg problem will have to be part of the development of new treatment protocols. In the future, there has to be a way to clearly define self and non-self. When that is achieved, we will see the successful treatment of cancers.

Chapter 12: NK cells and their role in the development of immune-based anti-cancer therapy

There are good reasons to include a chapter about NK cells in a book that focuses on T cell lymphocytes. Natural killer cells play a major role during the T cell response to developing tumors. Although they do not exclusively interact with the T cell, they do interact with the T cell immune system in an indirect but powerful way. Unlike the DC, which alters a T cell immune response through direct contact and communication, the NK cell does this without contact and sometimes at a great distance. Natural killer cells and T cells are related, but you would not know it until you examined surface features and cell surface protein expression. These similarities are just that: the two cells stay away from each other, but their long-distance relationship shapes the very nature of the T cell immune response to cancer. In Chapter 12 we attempt to explain NK cell immunobiology, which in some respects is a fledgling science. There are gaps and, quite frankly, mysteries surrounding exactly how these cells do what they do and why. We gather the available information and try to demonstrate the role of NK cells in the anti-cancer T cell response, as well as their use in immunotherapy.

The chapter begins with the conventional view of NK cells and introduces new information about the complexities of their function. For instance, NK cells have a series of receptors that work in a way opposite to the way T cell receptors work. When one receptor attaches to a normal MHC antigen, the signal it sends gets suppressed by signals from another receptor to inhibit NK cell activation and destruction of a normal cell. In order for the NK cell to kill a suspected cancer cell it must see an altered version (with glycoproteins in disorder) of an MHC class I protein or a specific protein associated with "cellular stress". The signals derived from these encounters override the inhibitory signals telling the NK cell not to kill. This inability to kill normal cells is the hallmark of NK cell function and makes them ideal candidates for anti-cancer immunotherapy. Even though some NK cells have similarities with T cells, such as a T cell receptor protein chain and the ability to secrete lymphokines, they are quite unique.

They play a pivotal role in the basic functions of the immune system and the anti-cancer T cell immune response. The normal immune responses to common pathogens depend heavily on the function of NK cells. This dependence is so acute that people born with birth defects associated with the lack of NK cells often die young from common infections by bacteria and viruses.

The premier function of an NK cell is its immune surveillance capability, where NK cells travel throughout the body, sampling cell surfaces for irregularities. Once these irregularities (usually associated with tumor development or viral/bacterial infections) are detected, the NK cell promptly kills that cell. The death of that cell sets off a series of events that involves an entire army of immune cells. This section describes how the NK cells recognize infected cells, the way they kill them, and the immune cells that are involved in the "clean-up" and distribution of information about the dead cells that the immune system needs in order to respond properly to the threat. The section describes where the NK cells go and what they do when they get there. Their influence over other cells of the immune system, such as T cells, B cells, and polymorphonuclear cells, is described as it relates to the everyday functioning of our immune system. This influence determines the type and intensity of a T cell immune response. There is a type of homeostatic relationship within the immune system that NK cells control and without it the immune system cannot function properly.

Natural killer cells have the ability to kill tumor cells using perforin or Fas-ligand. The immunological results of each type of killing affect the nature of the T cell response. A subset of NK cells also secrete lymphokines in the context of an anti-tumor T cell response. This function is a specialized one carried out by NK cells that appear at sites where there is either a viral/bacterial infection or a newly established tumor. The main lymphokine secreted during the response to a cancer is IL-2. Dendritic cells actually instruct NK cells not only to secrete IL2, but also to leave the draining lymph node (where this interaction occurs) and go to the site of the tumor. Once there, the NK cells secrete massive amounts of IL-2, which induces anti-tumor precursor CD8 T cells to become anti-tumor killer cells. They also secrete INF-g, which accelerates the anti-tumor response to diminish tumor cell numbers to the point where a simple or memory T cell response can finish them off. Natural killer cells cannot continue to secrete INF-g after the tumor is diminished. So a safeguard is built in. INF-g makes APCs present more tumor antigens by increasing the expression of MHC class I and class II proteins, which mediate this process. The safeguard comes in when there is continual exposure of APC to INF-g. Eventually, this exposure causes the APC to secrete IL-10 and TGF-B, which inhibit T cell activation and shut down NK cell activity, including the secretion of IFN-g.

The intercellular biochemical pathways of NK cells are just now being understood, 40 years after the discovery of the cells. The multiple redundant and intersecting pathways make identification of those that result in cellular changes difficult. However, some of these pathways are unique to the NK cell and demonstrate how this cell does what it does, by not changing into something else (differentiation), but just reacting. The predominant pathway pertains to the ability of NK cells to kill targets and secrete lymphokines. It is a direct line to NFkB activation, which can induce the DNA of the NK cell to turn on genes that are responsible for the activity mentioned above. This begins with the activation of a protein called caspase recruitment domain-9, or CARD9. The majority of the receptor systems in NK cell use this pathway, as opposed to only two obscure receptor systems in T and B cells. This pathway is important since it controls NK cell killing and lymphokine secretion, but also because of the use of NK cells as immunotherapeutic tool.

The unique properties of the NK cell can be exploited to treat cancer. One of the benefits of using NK cells in immunotherapy is that they are programmed, unlike T cell and B cells, to not attack normal tissues. This is related to the nature of their recognition of non-self. There is an inherent deterrent built into the design of their receptor system that allows these cells to recognize cells infected with viruses and or bacteria, and also newly formed cancer cells. Because of this they can suppress graft vs. host disease (GVHD) during bone marrow transplantation in cancer patients. This disease, caused by the T cells within the transplanted donor bone marrow cells, can be just as deadly to the recipient as the cancer that is being treated. The reason for this is complicated. Prior to traditional transplants, the donor and the recipient are MHC class I and II typed so that the cells do not respond to each other. The idea is for the T cells of the donor, once introduced into the recipient, to see the cancer as foreign and destroy it. The hitch: the donor's T cells can recognize the recipient's other tissues as foreign and attack them. Even if the donor and recipient were twins there are subtle changes in the DNA resulting in differences in tissue protein composition between the two individuals. The donor's T cells see this and attack tissues such as the skin, liver, and intestine, sometimes resulting in the death of the patient.

Immunologists attempted to reduce the incidence of GVDH by reducing the number of T cells in the donor's bone marrow. However, there was a different answer as to why GVHD was reduced when T cell-depleted donor bone marrow cells were used. The DCs of the recipient were the key. Since DCs are resistant to radiological and chemical elimination, prior to the transplant it was found that they expressed minor changes in proteins that the donor's T cells recognized, which induced the T cells to attack the recipient's organ systems. When scientists removed the T cells from the donor's bone marrow, they removed some NK cells, which share antigens with T cells. The T cell depletion method does not remove every single T cell, and new T cells are generated by the donor's transplanted bone marrow over time. This procedure left a sizable number of donor NK cells. The NK cells of the donor's bone marrow attacked the recipient's DCs. This, in effect, removed the cells that might be induced by any residual T cells to attack the recipient's organs. Therefore, the remaining donor NK cells stopped the generation of the T cell-mediated GVHD reaction.

Ideally, in a bone marrow transplant for treatment of leukemia, the donor and patient are matched for MHC class I and II antigens and the donor's T cells take up residence in the patient's body as if it was their own. However, the patient's leukemia is a different story. Often, a patient's system initially reacts to the leukemic cells but gets defeated. In the end, all that is left is the leukemia. When an identical donor's bone marrow is transplanted, those T cells have never seen that leukemic cell; they destroy the cells and leave a memory T cell response behind to protect the patient from a relapse. Again, this is in an ideal situation. Many times, the T cells of the donor take up permanent residence in the cancer patient's body, leaving them a T cell population that is mixed. This is called "chimerism" and is found in only one other species, the marmoset monkeys.

Immunologists studying graft vs. tumor or GVT were convinced that T cells that had developed (in the thymus of the recipient) from the donor's T cell-depleted bone marrow were responsible. The T cells that took up residence in the patient's body did do some of the killing, but what was eventually discovered caused immunologists to have second thoughts. It turns out that NK cells can see and kill cancer cells because they recognize a veritable panoply of altered cellular antigens. The NK cell receptors that see and induce activation are involved in "cellular stress". Cellular stress is brought about by the cell doing things that it was not designed to do, including unregulated growth. Two forces induce cellular stress: one, carcinogens, mutagens, and bio-agents that cause the cell to transform into a cancerous form; two, inherited genetic abnormalities such as the "Philadelphia chromosome". The receptors that recognize these antigens are powerful activators of NK cell killing. They instantly override inhibitory signals generated by other receptors.

During bone marrow transplant, there is a difference between the T cell's response and the NK cell's response to the differences in tissue peptides expressed by DCs in donor and recipient. The donor's NK cells can see these differences as well as the donor's T cells can. If the donor's T cells see them, they become primed, leave and find the nearest tissue expressing that peptide and do as much damage as possible to that tissue. In contrast, the NK cell sees this as an altered MHC class I or II protein and promptly kills the DC. The difference between the two responses has to do with cellular programming. The T cell response involves multiple changes in the cell until it attacks the target. The NK cell, on the other hand, simply sees the self-peptide associated with the MHC protein, realizes that it is foreign, and kills the cell.

Another and final aspect of how NK cells attenuate GVHD is at first hard to understand, but the logic is straightforward. The NKT cell expresses the TCR alpha chain. In a bone marrow transplant, many T cells are removed because there is not a perfect match with the leukemia patient or mismatched donor cells are used. The NKT cells are transferred with the donor's bone marrow. This is not a perfect match and the NKT cells become activated and kill donor T cells by the FasL/FasR system. Reduction of donor T cells slows or stops the development of GVHD. When activated, NKT cells also secrete large quantities of IL-4 and INF-g. When the two lymphokines are secreted together they induce a Th-2 type T cell response. This has to do with antibody secretion, but not the generation of killer T cells, which is a Th-1 T cell response. In effect, the NKT cells re-direct the T cell immune response to produce antibodies instead of CTL. The threatening T cells left over from the FasL killing are told to activate B cells, which may produce an antibody to a target tissue. This has happened, but the effects are much less deadly than the ones induced by organ-specific killer T cells.

Chapter 12 concludes with a description of the NK cell's unique function, composition, and role in immunity and in the anti-tumor T cell response. There are striking differences in NK and T cell function, yet both cells have a defined purpose in the immune system. Also discussed are recent findings in NK cell immunology about the ability of NK cells to kill their target and leave behind a chemo-attractant (CCL5) so the other immune cells can find the site. T cells release these attractants as well, but only after several rounds of T cell division. CCL5 expression is controlled by NK cell intercellular pathways that are unique to the NK cell. Natural killer cells can also be genetically modified to attack specific types of tumor cells. This procedure has been done using T cells, but often the inserted gene is lost because of the complex nature of T cell division and differentiation. Since NK cells generally do not have to divide and do not differentiate, stable genetically altered NK cells were able to kill a surprising number of tumor cells that expressed the target protein. Natural killer cells have the potential to become the premier anti-cancer immunotherapeutic cells. The DCs now hold that distinction, but they can become puppets of Tregs; many clinical trials using these cells have met with mixed results. The use of NK cells in anti-cancer immunotherapy is still in its infancy, but immunobiologists see the vast potential of these cells to treat cancer.

Chapter 13: How cancer cells influence the anti-cancer T cell immune response and why this is a problem during immunotherapeutic treatment of cancer

A growing cancer has an arsenal of intentional and unintentional weapons to trick, divert, or suppress the T cell response against it. A cancer cell behaving as a normal cell and expressing immune-related protein antigens can paralyze the T cell response long enough to give itself a foothold in the body. Cancer cells outnumber the T cells or deactivate them by inducing anti-self mechanisms. Events that occur during the initial phases of tumor growth can dictate the tempo of the T cell immune response and ultimately void any attempt by the T cell immune system to mount a successful response. Several "global" systematic effects that occur as a tumor grows need to be understood for immunotherapeutic treatments to be successful.

A large part of the chapter is devoted to ways in which an established tumor interferes with the anti-cancer T cell response and the body's normal functions. The interference with non-immune functions indirectly affects the entire immune system, which depends on the normal function of the body. Immunosuppression and the advent of opportunistic infections, caused by the cancer cells, divert resources from the T cell immune system that would otherwise be used to fight the cancer itself. These problems interfere with or inhibit the effectiveness of immune-based anti-cancer treatments. A tumor mass can induce inflammatory conditions that cause the creation of Tregs. Once Tregs are created, it may be virtually impossible for anti-tumor T cells to be created or get near the tumor cells.

Scientists are trying to improve the effectiveness of treatments by manipulating two aspects of immune cellular behavior and function. The first is blockage or activation of receptors or ligands found on T cells and DCs. The second is manipulation of intercellular pathways that direct T cells and DCs how to function. Once the pathways are activated by cancer-induced changes, the cells behave in a negative manner. The activation of inhibitory pathways can be altered by specific pathway inhibitors used on cells prior to their use in treating the cancer. The chapter ends with the possible benefits of using some of the improved immunotherapeutic protocols in cancer treatments.

Chapter 14: The T cell anti-cancer response, T cell regulation, and immunotherapy of neural crest cancers

Chapter 14 is a technical and complicated chapter and defines immunological terms. It highlights a challenging task that the T cell immune system must face when it responds to any cancer derived from cells of our nervous system: to remove a cancer derived from tissues that if they were normal would be deemed "untouchable" by the T cell immune system. This situation brings into focus the role of and problems caused by regulatory T cells (Tregs) before, during, and after an active T cell immune response to two neural crest cancers: glioblastoma multiforme and melanoma.

Both cancers elicit similar T cell responses but have different patient outcomes, and the events and conditions that cause this difference are discussed. An extensive evaluation of the role Tregs play in the response is included because of their negative impact on the effectiveness of immunotherapeutic treatments for these cancers. Finally, the immunotherapeutic approaches to these cancers are outlined. This acquaints the reader with how human clinical trials are performed and points out the problems encountered.

Chapter 15: The immunological hole: Will we fall in?

Why are some people immune and some not? During outbreaks of meningitis, for example, often people in close contact with the infected individual did not get infected. Meningitis is highly infective, yet many individuals are immune to the pathogens. When they get exposed to the agents, they do not even know it. Their immune systems silently remove the pathogens. On the other hand, some people have absolutely no immunity to the same pathogens. In other words, they have a hole in their immune system that is blind to these agents. This difference exists probably with respect to all immunogenic agents. Many diseases are induced by pathogens in our environment, but not every individual exposed to them suffers from the resulting immunological disease. Similarly, the incidence of cancer varies among individuals exposed to the same cancer-causing agents. The basis and effects of this variance are explained in Chapter 15.

The chapter briefly describes the immune cells involved and gives examples of how susceptibility to certain diseases can be explained by the immunological hole. It also describes how the T cell immune system is structured to cope with infectious agents and cancer. It shows how this structure is challenged by the immunological hole. It discusses how we view threatening diseases and the decisions we as a society make to deal with them. Finally, it presents an example of the hole and what it means for our survival as a species.

Concluding Remarks

What impact will successful cancer treatment have on our world? We can only speculate, since it will be some time before we eliminate cancer. Our society will begin to change. Individuals who otherwise would die from cancer will make major contributions to the advancement of our society. We have lost many people to cancer who could have changed our lives. Our world has been shaped by the absence of their contributions.

There is another more tangible loss, which has to do with economics. If just a fraction of the cancer deaths did not occur—2,000 to 3,000 people die each day from cancer—it would add billions of dollars to our economy. A good percentage of these people are hard working productive citizens. The health care costs for cancer patients—one of the most expensive—have been spiraling out of control for the past 20 years. They are right up there with heart health care in terms of the dollars spent each year. We just seem to absorb these economic consequences. If there were even 75% fewer cancer deaths per year, the savings in heath care cost and the resulting increase in economic productivity would probably pay off the current government budget deficit in three years.

Cancer also has a high personal cost. We all have lost brothers, sisters, mothers, fathers, other family members, and friends to this disease. To us these losses have nothing to do with the economy, contributions to society, or lost productivity. They create a void that will remain inside us forever. We miss the people we have lost and did not want them to suffer, but sometimes they did. We remember them as they were and block out that memory of the last few home or hospital visits. This is just a part of the emotional cost of cancer and in some respects the most meaningful. The feeling would have to be similar to losing a limb. Sometimes, like the limbless, we feel that they are still with us and perhaps they are, because the memories never go away. Some of us are never the same after the loss of a loved one from cancer. For years we can exist in a world where part of us is missing and cannot be found. We drift through our lives, trying to cope with the new reality that cancer has forced upon us. Over time, the way we feel changes or evolves into something that is just there in our minds.

These feelings are real, just as real as the existence of this disease. Removing cancer from our society would alleviate the suffering of not only the patient, but also of countless others who knew and perhaps loved this person. As a society we would be neglectful of the ones who died and their loved ones if we did not do everything we could to put an end to cancer. That end is near and only a coordinated effort by all of us can bring us to it.

It's time to put some things behind us and look toward the future. The past is quite valuable because it is a resource. This is one resource that we need not neglect. There is information out there that will ultimately lead to the end of cancer. This information is scattered. One article here, one article there and the scientific journals that publish cancer-related research are too numerous to mention. This work represents the work of individuals or groups of individuals scattered around the word. In some remote location a scientist sees something in the work of others and then has a new idea based on his reading. We, as scientists, rely heavily on the work of other scientists. Most scientists work alone and unseen in some lab on the seventh floor, but the world of cancer research is a web. It is a large interconnected labyrinth of ideas all oriented in the same direction.

There is a way to find out how to successfully treat cancer. Most of it is through hard work in the lab and the occasional new idea no scientist has ever had before. We live for those moments where an idea comes into our head with such incredible clarity that we already know how the experiment will turn out. Sometimes such progress come at a price of neglect of other things and even people. However, without original ideas science cannot move forward and we are doomed. The incidence of cancer in general has lessened, but the death rate of many forms of cancer, such as gliomas and melanoma, has not changed since the 1950s. When cancers like these are defeated we will truly be free of the "cancer death sentence".

Our society is made up of individuals. These individuals interact with each other in ways that shape the society we live in. Each person is part of a larger and somewhat intangible entity. That entity (society) needs everyone to grow and develop. The loss of its individual parts (people) from a disease like cancer invariably leaves gaps that would otherwise be filled with new ideas and major contributions. We rely on each other more than we think. When someone passes away, a chain of events is altered. Somewhere down the line someone unknown to the deceased is affected. This happens all around us every day, but we don't see it because in some respects it's a way of life. Yes, there are other reasons why individuals die, but just focusing on cancer is important. Cancer deaths can be stopped. In fact, fewer people die from cancer today than did 30 years ago, but too many still die from this disease. The types of cancer that remain untreatable are deadly and require new treatments not yet developed. In the future, the hope is that these treatments will be developed and we will live in a society where getting cancer means just getting treatment and not death.

This book makes an effort to explain what immunity to cancer is about. Many people just don't understand what happens inside their bodies when a normal cell turns into a cancerous one. The process of eliminating that cancerous cell(s) is the job of the immune system and specifically the job of the T cell immune system. Through a series of complicated steps, the T cell immune system usually eradicates a cancer without our even knowing it. We know that a cancer has formed only when the T cell immune system has failed to remove it. Only when you understand the level of complexity and sheer sophistication of the T cell immune system as it removes a cancer do you appreciate this incredible gift nature has bestowed upon us. Despite this incredible ability of the T cell immune system, it still needs help. This is where we come in. I'm not just talking about scientists, physicians, and laboratory personnel, but ordinary people like you and me. We have a responsibility to not only recognize the symptoms of cancer, but also do whatever we can as human beings to end this disease. The only way to do that is to understand it, and that's what this book is about. If this book causes just one person to either understand what a certain type of cancer treatment involves, or be inspired to pursue immunology as a career, it is worth the effort to write it. Maybe in the future, that one person will change the world and make it a much better place to live in.

References

Aarntzen EH, Figdor CG, Adema GJ, Punt CJ, de Vries IJ (2008) Dendritic cell vaccination and immune monitoring. Cancer Immunol Immunother 57: 1559–1568.

Ackermann S, Beckmann MW, Thiel F, Bogenrieder T (2007) Topotecan in cervical cancer. Int J Gynecol Cancer 17: 1215–23.

Aicher A, Shu GL, Magaletti D, Mulvania T, Pezzutto A, Craxton A, Clark EA (1999) Differential role for p38 mitogen-activated protein kinase in regulating CD40-induced gene expression in dendritic cells and B cells. J Immunol 163: 5786–5795.

Aisenberg AC (2006) Immunologic status of Hodgkin's disease. Cancer 19: 382–394.

Alexander MA, Bennicelli J, Guerry D (1989) Defective antigen presentation by human melanoma cell lines cultured from advanced, but not biologically early disease. J Immunol 142: 4070–4078.

Alter HJ, Conry-Cantilena C, Melpolder J, Tan D, Van Raden M, Herion D, Lau D, Hoofnagle JH (1997) Hepatitis C in asymptomatic blood donors. Hepatology 26: 29S–33S.

Alzabin S, Bhardwaj N, Kiefer F, Sawasdikosol S, Burakoff S (2009) Hematopoietic progenitor kinase 1 is a negative regulator of dendritic cell activation. J Immunol 182: 6187–6194.

Andre S, Tough DF, Lacroix-Desmazes S, Kaveri SV, Bayry J. (2009) Surveillance of antigen-presenting cells by CD4+ CD25+ regulatory T cells in autoimmunity: immunopathogenesis and therapeutic implications. Am J Pathol 174: 1575–1587.

Anirban R, Timothy W, Phares D, Koprowski HD, Hooper DC (2007) Failure to open the blood-brain barrier and deliver immune effectors to central nervous system tissues leads to the lethal outcome of silver-haired bat rabies virus infection. J Virol 81: 1110–1118.

Antony PA, Restifo N (2005) CD4+CD25+ T regulatory cells, immunotherapy of cancer, and IL-2. J Immunother 28: 120–128.

Baban B, Chandler PR, Sharma MD, Pihkala J, Koni PA, Munn DH, Mellor AL (2009) IDO activates regulatory T cells and blocks their conversion into Th17-like T cells. J Immunol 183: 2475–2483.

Bader P, Willasch A, Klingebiel T (2008) Monitoring of post-transplant remission of childhood malignancies: is there a standard? Bone Marrow Transplat 42: 31–34.

Ballard C (2009) AIDS and Other Epidemics. Gareth Stevens Publishing, Strongsville, OH.

Bannard O, Kraman M, Fearon D (2009) Pathways of memory CD8+ T cell development. Eur J of Immunol 39: 2083–2087.

Barton-Burke M, Dwinell DM, Kafkas L, Lavalley C, Sands H, Proctor C, Johnson E (2008) Graft-versus-host disease: a complex long-term side effect of cell transplantation. Oncology 22: 31–45.

Basta VP, Sherman PA, Ting JPY (1988) Detailed delineation of an interferon-G-responsive element important in human HLA-DRA gene expression in a glioblastoma multiform cell line. Proc Natl Acad Sci USA 85: 8618–8622.

Bayry J, Triebel F, Kaveri SV, Tough DF (2007) Human dendritic cells acquire a semimature phenotype and lymph node homing potential through interaction with CD4+CD25+ regulatory T cells. J Immunol 178: 4184–4193.

Beauchemin N, Draber P, Dveksler G, Gold P, Gray-Owen S, Grunert F, Hammarstroom S, Holmes KV, Karlsson A, Kuroki M, Lin SH, Lucka L, Najjar SM, Neumaier M, Obrink B, Shively JE, Skubitz KM, Stanners CP, Thomas P, Thompson JA, Virji M, von Kleist S, Wagener C, Watt S, Zimmermann W (1999) Redefined nomenclature for members of the carcinoembryonic antigen family. Exp Cell Res 252: 243–249.

Beck KE, Blansfield JA, Tran KQ (2006) Enterocolitis in patients with cancer after antibody blockade of cytotoxic T-lymphocyte–associated antigen 4. J Clin Oncol 24: 2283–2289.

Bercovici N, Haicheur N, Massicard S, Vernel-Pauillac F, Adotevi O, Landais D, Gorin I, Robert C, Prince HM, Grob JJ, Leccia MT, Lesimple T, Wijdenes J, Bartholeyns J, Fridman WH, Salcedo M, Ferries E, Tartour E (2008) Analysis and characterization of anti-tumor T cell response after administration of dendritic cells loaded with allogeneic tumor lysate to metastatic melanoma patients. J Immunother 31: 101–112.

Bergwelt-Baildon M, Popov A, Saric T, Cheminitz J, Classen S, Stoffel MS, Fiore F, Roth U, Beyer M, Debey S, Wickenhauser C, Hanish G, Schultze J (2006) CD25 and indoleamine 2,3-deoxygenase are up-regulated by prostaglandin E2 and expressed by tumor-associated dendritic cells in vivo: additional mechanisms of T-cell inhibition. Blood 108: 228–237.

Bertaina A, Locatelli F, Moretta L (2009) Transplantation and innate immunity: the lesson of natural killer cells. Ital J Pediatr 35: 44–46.

Berzofsky JA, Terabe M (2009) The contrasting roles of NKT cells in tumor immunity. Curr Mol Med 6: 667–672.

Birmingham K (2002) What is translational research? Nat Med 8: 647–649.

Bluestone JA, Abbas AK (2003) Natural versus adaptive regulatory T cells. Nat Rev Immunol 3: 253–257.

Bluestone JA, Mackay CR, O'Shea JJ, Stockinger B (2009) Viewpoint: The functional plasticity of T cell subsets. Nat Rev Immunol 9: 811–816.

Bohle A (2010) Bladder cancer: meta-analysis of BCG versus mitomycin C—a deeper insight? Nat Rev Urol 7: 8–10.

Boissonnas A, Scholer-Dahirel A, Simon-Blancal V, Pace L, Valet F, Kissenpfennig A, Sparwasser T, Malissen B, Fetler L, Amigorena S (2010) Foxp3+ T cells induce perforin-dependent dendritic cell death in tumor-draining lymph nodes. Immunity 32: 266–278.

Bonifaz LC, Bonnyay DP, Charalambous A, Darguste DI, Fujii S, Soares H, Brimnes MK, Moltedo B, Moran TM, Steinman RM (2004) In vivo targeting of antigens to maturing dendritic cells via the DEC-205 receptor improves T cell vaccination. J Exp Med 199: 815–824.

Bopp T, Radsak M, Schmitt E, Schild H (2010) New strategies for the manipulation of adaptive immune responses. Cancer Immunol Immunother. 2010 Apr 2. [Epub ahead of print]

Bozzacco L, Trumpfheller C, Siegal FP, Mehandru S, Markowitz M, Carrington M, Nussenzweig MC, Piperno AG, Steinman RM (2007) DEC-205 receptor on dendritic cells mediates presentation of HIV gag protein to CD8+ T cells in a spectrum of human MHC I haplotypes. Proc Natl Acad Sci USA 104: 1289–1294.

Brooks WH, Markesbery WR, Gupta GD, Roszman TL (1978) Relationship of lymphocyte invasion and survival of brain tumor patients. Ann Neurol 4: 219–224.

Burnet FM (1971) Immunological surveillance in neoplasia. Transplant Rev 7: 3–25.

Busch R, Doebele RC, von Scheven E, Fahrni J, Mellins ED (1998) Aberrant intermolecular disulfide bonding in a mutant HLA-DM molecule: implications for assembly, maturation, and function. J Immunol 160: 734–743.

Caligiuri MA (2008) Human natural killer cells. Blood 112: 461–469.

Carrega P, Pezzino G, Queirolo P, Bonaccorsi I, Falco M, Vita G, Pende D, Misefari A, Moretta A, Mingari MC, Moretta L, Ferlazzo G (2009) Susceptibility of human melanoma cells to autologous natural killer (NK) cell killing: HLA-related effector mechanisms and role of unlicensed NK cells. PLoS One 4: e8132.

Cassatella MA (2003) The neutrophil: an emerging regulator of inflammatory and immune response. S. Karger AC, Basel, Switzerland.

Chanarin I (1989) Laboratory Haematology: An Account of Laboratory Techniques. Churchill Livingstone, Elsever Heath Sciences Division, Philadelphia, PA.

Chattopadhyay S, Chakraborty NG, Mukherji B (2005) Regulatory T cells and tumor immunity. Cancer Immunol Immunother 54: 1153–1161.

Chen W, Liang X, Peterson AJ, Munn DH, Blazar BR (2008) The indoleamine 2,3-dioxygenase pathway is essential for human plasmacytoid dendritic cell-induced adaptive T regulatory cell generation. J Immunol 181: 5396–5404.

Cheong SK, Eow GI, Leong CF (2002) Non-myeloablative conditioning for hemopoietic stem cell transplantation—does it work? Malays J Pathol 24: 1–8.

Cho JH, Kim HO, Surh CD, Sprent J (2010) T cell receptor-dependent regulation of lipid lafts controls naive CD8(+) T cell homeostasis. Immunity. [Epub ahead of print]

Choi J, Ritchey J, Prior JL, Holt M, Shannon WD, Deych E, Piwnica-Worms DR, Dipersio JF (2010) In vivo administration of hypomethylating agents mitigate graft-versus-host disease (GvHD) without sacrificing graft-versus-leukemia (GvL). Blood. 2010 Apr 27. [Epub ahead of print]

Cifuentes M, Albala C, Rojas CV (2008) Differences in lipogenesis and lipolysis in obese and non-obese adult human adipocytes. Biol Res 41: 197–204.

Clark RA, Adinoff AD (1989) Aeroallergen contact can exacerbate atopic dermatitis: patch tests as a diagnostic tool. J Am Acad Dermatol 21: 863–869.

Clark WH, Elder DE, Guerry D (1984) A study of tumor progression: the precursor lesions of superficial spreading and nodular melanoma. Hum Pathol 15: 1147–1165.

Clark WH, Elder DE, Van Horn M (1986) The biologic forms of malignant melanoma. Hum Pathol 17: 443–450.

Cohen S (1990) Lymphokines and the Immune Response. CRC Press, Inc., Boca Raton, FL.

Cong Y, Konrad A, Iqbal N, Hatton RD, Weaver CT, Elson CO (2005) Generation of antigen-specific, Foxp3-expressing CD4+ regulatory T cells by inhibition of APC proteosome function. J Immunol 174: 2787–2795.

Corinti S, Albanesi C, la Sala A, Pastore S, Girolomoni G (2001) Regulatory activity of autocrine IL-10 on dendritic cell functions. J Immunol 166: 4312–4318.

Culver KW (1996) Gene therapy for malignant neoplasms of the CNS. Bone Marrow Transplant 18: 6–9.

Curiel TJ (2007) Tregs and rethinking cancer immunotherapy. J Clin Invest 117: 1167–1174.

Curtin JF, Candolfi M, Fakhouri TM (2008) Treg depletion inhibits efficacy of cancer immuno-therapy: implications for clinical trials. PLosONE 3: 1–17.

Dannull J, Su Z, Rizzieri D, Yang BK, Coleman D, Yancey D, Zhang A, Dahm P, Chao N, Gilboa E, Vieweg J (2005) Enhancement of vaccine-mediated antitumor immunity in cancer patients after depletion of regulatory T cells. J Clin Invest 115: 3623–3633.

Dao T, Scheinberg DA (2008) Peptide vaccines for myeloid leukaemias. Best Pract Res Clin Haematol 21: 391–404.

Dauer M, Schnurr M, Eigler A (2008) Dendritic cell-based cancer vaccination: quo vadis? Expert Rev Vaccines 7: 1041–1053.

Davis D (2007) The Secret History of the War on Cancer. Perseus Book Group, Philadelphia, PA.

De Vleeschouwer S, Fieuws S, Rutkowski S, Van Calenbergh F, Van Loon J, Sciot R, Wilms G, Demaelrel P, Warmuth-mentz M, Soerensen N, Wolff JE, Wagner S, Kaempgen E, Van Gool SW (2008) Postoperative adjuvant dendritic cell-based immunotherapy in patients with relapsed glioblastoma multiforme. Clin Cancer Res 14: 30098–30104.

de Witte MA, Jorritsma A, Kaiser A, van den Boom MD, Dokter M, Bendle GM, Haanen JB, Schumacher TN (2008) Requirements for effective antitumor responses of TCR transduced T cells. J Immunol 181: 5128–5136.

Doherty PC, Kelso A (2008) Toward a broadly protective influenza vaccine. J Clin Invest 118: 3273–3275.

Drake CG, Jaffee E, Pardoll DM (2006) Mechanisms of immune evasion by tumors. Adv Immunol 90: 51–81.

Dudley ME, Yang JC, Sherry R, Hughes MS, Royal R, Kammula U, Robbins PF, Huang J, Citrin DE, Leitman SF, Wunderlich J, Restifo NP, Thomasian A, Downey SG, Smith FO, Klapper J, Morton K, Laurencot C, White DE, Rosenberg SA (2008) Adoptive cell therapy for patients with metastatic melanoma: evaluation of intensive myeloablative chemoradiation preparative regimens. J Clin Oncol 26: 5233–5339.

Dumitriu IE, Dunbar DR, Howie SE, Sethi T, Gregory CD (2009) Human dendritic cells produce TGF-beta 1 under the influence of lung carcinoma cells and prime the differentiation of CD4+CD25+Foxp3+ regulatory T cells. J Immunol 182: 2795–2807.

Edwards LJ, Evavold BD (2010) A unique unresponsive CD4+ T cell phenotype post TCR antagonism. Cell Immunol 261: 64–68.

El Andaloussi A, Lesniak MS (2006) An increase in CD4+CD25+FOXP3+ regulatory T cells in tumor-infiltrating lymphocytes of human glioblastoma multiforme. Neuro Oncol 8: 234–243.

Elgert KD (2009) Immunology: Understanding the Immune System, 2nd edition. Wiley, Hoboken, NJ.

Enzinger PC, Ilson DH, Kelsen DP (1999) Chemotherapy in esophageal cancer. Semin Oncol 2: 12–20.

Fearon DT, Carr JM, Telaranta C, Carrasco MJ, Thaventhiran JE (2006) The rational for the IL-2-independent generation of self-renewing central memory CD8+ T cells. Immunol Rev 211: 104–118.

Feldmann M, Lamb JR, Owen MJ (1989) T Cells. Wiley, Hoboken, NJ.

Ferlazzo G, Munz C (2009) Dendritic cell interactions with NK cells from different tissues. J Clin Immunol 29: 265–273.

Field WH, Steck, DJ, Smith BJ, Brus CP, Fisher EL, Neuberger JS, Platz CE, Robinson RA, Woolson RF, Lynch CF (2000) Residential Radon gas exposure and lung cancer: The Iowa radon lung cancer study. Am J Epidemiol 151: 1091–1102.

Fontenot JD, Rudensky AY (2005) A well adapted regulatory contrivance: regulatory T cell development and the forkhead family transcription factor Foxp3. Nat Immunol 6: 331–337.

Frank NY, Schatton T, Frank MH (2010) The therapeutic promise of the cancer stem cell concept. J Clin Invest 120: 41–50.

Fujimoto MA, Kristedja A, Bhachu TS, Ye S, Deck X, O'Day SJ (2006) Enhanced survival associated with vitiligo expression during maintenance biotherapy for melastatic melanoma. J Invest Dermatol 126: 2658–2663.

Gabrilovich DI, Chen HL, Girgis KR, Cunningham HT, Meny GM, Nadaf S, Kavanaugh D, Carbone DP (1996) Production of vascular endothelial growth factor by human tumors inhibits the functional maturation of dendritic cells. Nat Med. 2: 1096–1103.

Galgani M, Di Giacomo A, Matarese G, La Cava A (2009) The Yin and Yang of CD4(+) regulatory T cells in autoimmunity and cancer. Curr Med Chem 16: 4626–4631.

Gallia G, Houff SA, Major EO, Kalili K (1997) Review: JC virus infection of lymphocytes—revisited. J Infec Diseases 176:1603–1609.

Garrod KR, Wei SH, Parker I, Cahalan MD (2007) Natural killer cells actively patrol peripheral lymph nodes forming stable conjugates to eliminate MHC-mismatched targets. Proc Natl Acad Sci USA 104: 12081–12086.

Gattinoni L, Finkelstein SE, Klebanoff CA, Antony PA, Palmer DC, Spiess PJ, Hwang LN, Yu Z, Wrzesinski C, Heimann DM (2005) Removal of homeostatic cytokine sinks by lymphodepletion enhances the efficacy of adoptively transferred tumor-specific $CD8^+$ T cells. J Exp Med 202: 907–912.

Getnet D, Maris CH, Hipkiss EL, Grosso JF, Harris TJ, Yen HR, Bruno TC, Wada S, Adler A, Georgantas RW, Jie C, Goldberg MV, Pardoll DM, Drake CG (2009) Tumor recognition and self-recognition induce distinct transcriptional profiles in antigen-specific CD4 T cells. J Immunol 182: 4675–4685.

Ghosh N, Matsui W (2009) Cancer stem cells in multiple myeloma. Cancer Lett 277: 1–7.

Ghosh TK, Mickelson DJ, Fink J, Solberg JC, Inglefield JR, Hook D, Gupta SK, Gibson S, Alkan SS (2006) Toll-like receptor (TLR) 2–9 agonists-induced cytokines and chemokines: I. Comparison with T cell receptor-induced responses. Cell Immunol 243: 48–57.

Glynn SA, Boersma BJ, Howe TM, Edvardsen H, Geisler SB, Goodman JE, Ridnour LA, Lonning PE, Borresen-Dale AL, Naume B, Kristensen VN, Chanock SJ, Wink DA, Ambs S (2009) A mitochondrial target sequence polymorphism in manganese superoxide dismutase predicts inferior survival in breast cancer patients treated with cyclophosphamide. Clin Cancer Res 15: 4165–4173.

Gorski A, Krotkiewski H, Zimecki M (2001) Autoimmunity. Springer Publishing, New York, NY.

Grauer OM, Nierkens S, Bennink E, Toonen LW, Boon L, Wesseling P, Sutmuller RP, Adema GJ (2007) CD4+FoxP3+ regulatory T cells gradually accumulate in gliomas during tumor growth and efficiently suppress antiglioma immune responses in vivo. Int J Cancer 121: 95–105.

Grauer OM, Sutmuller RP, van Maren W, Jacobs JF, Bennink E, Toonen LW, Nierkens S, Adema GJ (2008) Elimination of regulatory T cells is essential for an effective vaccination with tumor lysate-pulsed dendritic cells in a murine glioma model. Int J Cancer 122: 1794–1802.

Grauer OM, Molling JW, Bennink E, Toonen LW, Sutmuller RP, Nierkens S, Adema GJ (2008a) TLR ligands in the local treatment of established intracerebral murine gliomas. J Immunol 181: 6720–6729.

Groschel S, Piggott KD, Vaglio A, Ma-Krupa W, Singh K, Goronzy JJ, Weyand CM (2008) TLR-mediated induction of negative regulatory ligands on dendritic cells. J Mol Med 86: 443–455.

Guerry D, Alexander MA, Elder DE, Herlyn MF (1987) Interferon-gamma regulates the T cell response to precursor nevi and biologically early melanoma. J Immunol 139: 305–312.

Guerry D, Alexander MA, Herlyn MF, Zehngebot LM, Mitchell KF, Zmijewski CM, Lusk EJ (1984) HLA-DR histocompatibility leukocyte antigens permit cultured human melanoma cells from early but not advanced disease to stimulate autologous lymphocytes. J Clin Invest 73: 267–271.

Gui U, Killic A, Tulunay O, Kaygusuz G (2007) Vitiligo associated with malignant melanoma and lupus erythematosis. J Dermatol 34: 142–145.

Hanig J, Lutz MB (2008) Suppression of mature dendritic cell function by regulatory T cells in vivo is abrogated by CD40 licensing. J Immunol 180: 1405–1413.

Hara H, Ishihara C, Takeuchi A, Xue L, Morris SW, Penninger JM, Yoshida H, Saito T (2008) Cell type-specific regulation of ITAM-mediated NF-kappaB activation by the adaptors, CARMA1 and CARD9. J Immunol 181: 918–930.

Haraguchi K, Takahashi T, Matsumoto A, Asai T, Kanda Y, Kurokawa M, Ogawa S, Oda H, Taniguchi M, Hirai H, Chiba S (2005) Host-residual invariant NK T cells attenuate graft-versus-host immunity. J Immunol 175: 1320–1328.

Heemskerk B, Liu K, Dudley ME, Johnson LA, Kaiser A, Downey S, Zheng Z, Shelton TE, Matsuda K, Robbins PF, Morgan RA, Rosenberg SA (2008) Adoptive cell therapy for patients with melanoma, using tumor-infiltrating lymphocytes genetically engineered to secrete interleukin-2. Hum Gene Ther 19: 496–510.

Hersey P, Halliday GM, Farrelly ML (2008) Phase I/II study of treatment with matured dendritic cells with or without low dose IL-2 in patients with disseminated melanoma. Can Immunol Immunother 57: 1039–1051.

Higgins SC, Lavelle EC, McCann C, Keogh B, McNeela E, Byrne P, O'Gorman B, Jarnicki A, McGuirk P, Mills KH (2003) Toll-like receptor 4-mediated innate IL-10 activates antigen-specific regulatory T cells and confers resistance to Bordetella pertussis by inhibiting inflammatory pathology. J Immunol 171: 3119–3127.

Holmberg D, Cilio CM, Lundholm M (2005) CTLA-4 (CD152) and its involvement in autoimmune disease. Autoimmunity 38: 225–233.

Holmes S, He M, Xu T, Lee PP (2005) Memory T cells have gene expression patterns intermediate between naïve and effector. Proc Natl Acad Sci USA 102: 5519–5523.

Horn J, Wang X, Reichardt P, Stradal TE, Warnecke N, Simeoni L, Gunzer M, Yablonski D, Schraven B, Kliche S (2009) Src homology 2-domain containing leukocyte-specific phosphoprotein of 76 kDa is mandatory for TCR-mediated inside-out signaling, but dispensable for CXCR4-mediated LFA-1 activation, adhesion, and migration of T cells. J Immunol 183: 5756–5767.

Hung K, Hayashi R, Lafond-Walker A, Lowenstein C, Pardoll D, Levitsky H. (1998) The central role of CD4(+) T cells in the antitumor immune response. J Exp Med 188: 2357–2368.

Igarashi T, Wynberg J, Srinivasan R, Becknell B, McCoy JP Jr, Takahashi Y, Suffredini DA, Linehan WM, Caligiuri MA, Childs RW (2004) Enhanced cytotoxicity of allogeneic NK cells with killer immunoglobulin-like receptor ligand incompatibility against melanoma and renal cell carcinoma cells. Blood 104: 170–177.

Iwamoto T, Kagawa Y, Naito Y, Kuzuhara S, Kojima M (2004) Steroid-induced diabetes mellitus and related risk factors in patients with neurologic diseases. Pharmacotherapy 24: 508–514.

Izumoto S, Tsuboi A, Oka Y (2008) Phase II clinical trail of Whilms tumor 1 peptide vaccination in patients with recurrent glioblastoma multiforme. J Neurosurg 108: 963–971.

James WD, Odom RB (1983) Graft-v-host disease. Arch Dermatol 119: 683–689. Jandus C, Bioley G, Speiser DE, Romero P (2008) Selective accumulation of differentiated FOXP3(+) CD4 (+) T cells in metastatic tumor lesions from melanoma patients compared to peripheral blood. Cancer Immunol Immunother 57: 1795–1805.

Jarnicki AG, Conroy H, Brereton C, Donnelly G, Toomey D, Walsh K, Sweeney C, Leavy O, Fletcher J, Lavelle EC, Dunne P, Mills KH (2008) Attenuating regulatory T cell induction by TLR agonists through inhibition of p38 MAPK signaling in dendritic cells enhances their efficacy as vaccine adjuvants and cancer immunotherapeutics. J Immunol 180: 3797–3806.

Jonuleit H, Schmitt E, Schuler G, Knop J, Enk AH (2000) Induction of interleukin 10-producing, nonproliferating CD4(+) T cells with regulatory properties by repetitive stimulation with allogeneic immature human dendritic cells. J Exp Med 192: 1213–1222.

Jordan KR, McMahan RH, Kemmler CB, Kappler JW, Slansk JE (2010) Peptide vaccines prevent tumor growth by activating T cells that respond to native tumor antigens. Proc Natl Acad Sci USA, In Press.

Joshi AD, Clark EM, Wang P, Munger CM, Hegde GV, Sanderson S, Dave HP, Joshi SS (2007) Immunotherapy of human neuroblastoma using umbilical cord blood-derived effector cells. J Neuroimmune Pharmacol 2: 202–212.

June CH (2007) Principals of adoptive T cell cancer therapy. J Clin Inves 117: 1204–1212.

Juszczak M (2009) Th17 cells in the pathogenesis of multiple sclerosis. Postepy Hig Med Dosw (Online) 63: 492–501.

Kalinski P, Urban J, Narang R, Berk E, Wieckowski E, Muthuswamy R (2009) Dendritic cell-based therapeutic cancer vaccines: what we have and what we need. Future Oncol 5: 379–390.

Kalinski P, Wieckowski E, Muthuswamy R, de Jong E (2010) Generation of stable Th1/CTL-, Th2-, and Th17-inducing human dendritic cells. Methods Mol Biol 595: 117–133.

Kamala T, Nanda K (2009) Protective response to Leishmania major in BALB/c mice requires antigen processing in the absence of DM. J Immunol 182: 4882–4890.

Kaplan DE, Ikeda F, Li Y, Nakamoto N, Ganesan S, Valiga ME, Nunes FA, Rajender Reddy K, Chang KM (2008) Peripheral virus-specific T-cell interleukin-10 responses develop early in acute hepatitis C infection and become dominant in chronic hepatitis. J Hepatol 48: 903–913.

Katz JD, Bonavida B (1989) A spontaneous sarcoma dependent on host tumor-specific immune lymphocytes. Bioessays 11: 181–185.

Kernan NA, DuPont B (1996) Minor histocompatibility antigens and marrow transplantation. NEJM 334: 323–324.

Kiessling AA, Anderson S (2003) Human Embryonic Stem Cells. Jones and Bartlett Publishers, Sudbury, MA.

Kim JH, Choi EY, Chung DH (2007) Donor bone marrow type II (non-Va14Ja18 CD1d-restricted) NKT cells suppress Graft-versus-host disease by producing INF-g and IL-4. J Immunol 178: 6579–6587.

Kim JY, Kang JS, Kim HM, Kim YK, Lee HK, Song S, Hong JT, Kim Y, Han SB (2009) Inhibition of phenotypic and functional maturation of dendritic cells by manassantin a. J Pharmacol Sci 109: 583–592.

Kirsch M, Black P M (2004) Angiogenesis in Brain Tumors. Kluwer Academic Publishers, Norwell, MA.

Klein G, Vande Woude GF (2007) Advances in Cancer Research. Elsevier, Maryland Heights, MO.

Koike K, Takaki A, Shiretori Y (2004) Correlation between the grade of immune response against hepatitis C virus (HCV) infection and MHC class II polymorphisms. Nippon Rinsho 62: 147–150.

Kong LY, Wei J, Sharma AK, Barr J, Abou-Ghazal MK, Fokt I, Weinberg J, Rao G, Grimm E, Priebe W, Heimberger AB (2009) A novel phosphorylated STAT3 inhibitor enhances T cell cytotoxicity against melanoma through inhibition of regulatory T cells. Cancer Immunol Immunother 58: 1023–1032.

Korn T, Bettelli E, Oukka M, Kuchroo VK (2009) IL-17 and Th17 Cells. Annu Rev Immunol 27: 485–517.

Kornstein MJ, Brooks JSJ, Elder DE (1983) Immunperoxidase localization of lymphocyte subsets in the host response to melanoma and nevi. Cancer Res 43: 2749–2755.

Kronenberg M (2005) Toward an understanding of NKT cell biology: progress and paradoxes. Annu Rev Immunol 23: 877–900.

Krummel MF, Macara I (2006) Maintenance and modulation of T cell polarity. Nat Immunol 7: 1143–1149.

Kumar D, Hosse J, von Toerne C, Noessner E, Nelson PJ (2009) JNK MAPK pathway regulates constitutive transcription of CCL5 by human NK cells through SP1. J Immunol 182: 1011–1020.

Lau LL, Jiang J, Shen H (2005) In vivo modulation of T cell responses and protective immunity by TCR antagonism during infection. J Immunol 174: 7970–7976.

Lazovic J, Jensen MC, Ferkassian E, Aquilar B, Raubitscheck A, Jacobs RE (2008) Imaging immune response in vitro: cytotoxic action of genetically altered T cells directed to glioblastoma multiforme. Clin Cancer Res 14: 3832–3839.

Lefranc M, Lefranc G (2001) The T cell Receptor Factsbook. Academic Press, New York, NY.

Le Gal FA, Ayyoub M, Dutoit V, Widmer V, Jager E, Cerottini JC, Dietrich PY, Valmori D (2005) Distinct structural TCR repertoires in naturally occurring versus vaccine-induced CD8 T-cell responses to the tumor specific antigen NY-ESO-1. J Immunother 28: 252–257.

Lepisto AJ, Moser AJ, Zeh H, Lee K, Bartlett D, McKolanis JR, Geller BA, Schmotzer A, Potter DP, Whiteside T, Finn OJ, Ramanathan RK (2008) A phase I/II study of a MUC1 peptide pulsed autologous dendritic cell vaccine as adjuvant therapy in patients with resected pancreatic and biliary tumors. Cancer Ther 6: 955–964.

Lesterhuis WJ, Aarntzen EH, De Vries IJ (2008) Dendritic cell vaccines in melanoma: from promise to proof. Crit Rev Oncol Hematol 66: 118–134.

Levine TP, Chain BM (1991) The cell biology of antigen processing. Crit Rev Biochem Mol Biol 26: 439–473.

Liang B, Workman C, Lee J, Chew C, Dale BM, Colonna L, Flores M, Li N, Schweighoffer E, Greenberg S, Tybulewicz V, Vignali D, Clynes R (2008) Regulatory T cells inhibit dendritic cells by lymphocyte activation gene-3 engagement of MHC class II. J Immunol 180: 5916–5926.

Liu VC, Wong LY, Jang T, Shah AH, Park I, Yang X, Zhang Q, Lonning S, Teicher BA, Lee C (2007) Tumor evasion of the immune system by converting CD4+CD25- T cells into CD4+CD25+ T regulatory cells: role of tumor-derived TGF-beta. J Immunol 178: 2883–2892.

Liu X, Newton RC, Friedman SM, Scherle PA (2009a) Indoleamine 2,3-dioxygenase, an emerging target for anti-cancer therapy. Curr Cancer Drug Targets 9: 938–952.

Liu Z, Kim JH, Falo LD, You Z (2009b) Tumor regulatory T cells potently abrogate antitumor immunity. J Immunol 182: 6160–6167.

Lombardi G, Luciano A, Riffo-Vasquez Y (2009a) Dendritic Cells. Springer-Verlag, Berlin, Heidelberg, GBM.

Lombardi V, Van Overtvelt L, Horiot S, Moingeon P (2009) Human dendritic cells stimulated via TLR7 and/or TLR8 induce the sequential production of Il-10, IFN-gamma, and IL-17A by naive CD4+ T cells. J Immunol 182: 3372–3379.

Long SA, Walker MR, Rieck M, James E, Kwok WW, Sanda S, Pihoker C, Greenbaum C, Nepom GT, Buckner JH (2009) Functional islet-specific Treg can be generated from CD4+CD25- T cells of healthy and type 1 diabetic subjects. Eur J Immunol 39: 612–620.

Longhi MP, Wright K, Lauder SN, Nowell MA, Jones GW, Godkin AJ, Jones SA, Gallimore AM (2008) Interleukin-6 is crucial for recall of influenza-specific memory CD4 T cells. PLoS Pathog 4: e1000006.

Loose D, Van de Wiele C (2009) The immune system and cancer. Cancer Biother Radiopharm 24: 369–376.

Lundqvist A, McCoy JP, Samsel L, Childs R (2007) Reduction of GVHD and enhanced antitumor effects after adoptive infusion of alloreactive Ly49-mismatched NK cells from MHC-matched donors. Blood 109: 3603–3606.

Luptrawan A, Liu G, Yu JS (2008) Dendritic cell therapy for malignant gliomas. Rev Recent Clin Trials 3: 10–21.

Ma DY, Clark EA (2009) The role of CD40 and CD154/CD40L in dendritic cells. Semin Immunol 21: 265–272.

Mahnke K, Johnson TS, Ring S, Enk, AH (2007) Tolerogenic dendritic cells and regulatory T cells: a two-way relationship. J Dermatol Sci 46: 159–67.

Marguti I, Yamamoto GL, da Costa TB, Rizzo LV, de Moraes LV (2009) Expansion of CD4+ CD25+ Foxp3+ T cells by bone marrow-derived dendritic cells. Immunology 127: 50–61.

Marin-Esteban V, Falk K, Rotzschke O (2004) "Chemical analogues" of HLA-DM can induce a peptide-receptive state in HLA-DR molecules. J Biochem 239: 50684–50690.

Marks F, Klingmuller U, Muller-Decker K (2009) Cellular Signal Processing: An Introduction to the Molecular Mechanisms of Signal Transduction. Garland Science, New York, NY.

Marusina AI, Burgess SJ, Pathmanathan I, Borrego F, Coligan JE (2008) Regulation of human DAP10 gene expression in NK and T cells by Ap-1 transcription factors. J Immunol 180: 409–417.

Matsushima H, Takashima A (2010) Bidirectional homing of Tregs between the skin and lymph nodes. J Clin Invest 120: 653–656.

McLendon RE, Halperin EC (2003) Is the long-term survival of patients with intracranial glioblastoma multiforme overstated? Cancer 98: 1745–1748.

Melillo JA, Song L, Bhagat G, Blazquez AB, Plumlee CR, Lee C, Berin C, Reizis B, Schindler C (2010) Dendritic cell (DC)-specific targeting reveals Stat3 as a negative regulator of DC function. J Immunol 184: 2638–2645.

Merad M, Ginhoux F, Collin M (2008) Origin, homeostasis and function of Langerhans cells and other langerin-expressing dendritic cells. Nat Rev Immunol 8: 935–947.

Merad M, Manz MG (2009) Dendritic cell homeostasis. Blood 113: 3418–27.
Moretta L (2003) NK Cells: Origin, Receptors and Specificity. Landes Bioscience, Austin, TX.
Morgan RA, Dudley ME, Wunderlich JR, Hughes MS, Yang JC, Sherry RM, Royal RE, Topalian SL, Kammula US, Restifo NP, Zheng Z, Nahvi A, de Vries CR, Rogers-Freezer LJ, Mavroukakis SA, Rosenberg SA (2006) Cancer regression in patients after transfer of genetically engineered lymphocytes. Science 314: 126–129.
Mottet C, Uhlig HH, Powrie F (2003) Cutting edge: cure of colitis by CD4+CD25+ regulatory T cells. J Immunol 170: 3939–3943.
Mougiakakos D, Choudhury A, Lladser A, Kiessling R, Johansson CC (2010) Regulatory T cells in cancer. Adv Cancer Res 107: 57–117.
Munn DH, Sharma MD, Mellor AL (2004) Ligation of B7-1/B7-2 by human CD4+ T cells triggers indoleamine 2,3-dioxygenase activity in dendritic cells. J Immunol 172: 4100–4110.
Nagata S (1996) A death factor—the other side of the coin. Behring Inst Mitt 97: 1–11.
Navratil JS, Sabatine JM, Ahearn JM (2004) Apoptosis and immune responses to self. Rheum Dis Clin North Am 30(1): 193–212.
Nchinda G, Kuroiwa J, Oks M, Trumpfheller C, Park CG, Huang Y, Hannaman D, Schlesinger SJ, Mizenina O, Nussenzweig MC, Uberla K, Steinman RH (2008) The efficacy of DNA vaccination is enhanced in mice by targeting the encoded protein to dendritic cells. J Clin Invest 118: 1427–1436.
Neumann J, Eis-Hubinger AM, Koch N (2003) Herpes simplex virus type 1 targets the MHC class II processing pathway for immune evasion. J Immunol 171: 3071–3085.
Neumann J, Koch N (2005) Assembly of major histocompatibility complex class II subunits with invariant chain.FEBS Lett 579: 6055–6059.
Norris PJ, Stone JD, Anikeeva N (2006) Antagonism of HIV-specific CD4 T cells by C-terminal truncation of a minimum epitope. Mol Immunol 43: 1349–1357.
Nowell PC, Emanuel BS, Finan JB, Erikson J, Croce CM (1984) Chromosome rearrangements in oncogenesis. Microbiol Sci 1: 223–228.
Obar JJ, Lefranois L (2010) Memory CD8+ T cell differentiation. Ann NY Acad Sci 1183: 251–266.
O' Hagan, DT (2000) Vaccine Adjuvants. Preparation Methods and Research Protocols. Humana Press, Totowa, NJ.
Ohgaki H, Kleihues P (2005) Epidemiology and etiology of gliomas. Acta Neuropathol 109: 93–108.
Okada H, Lieberman FS, Walter KA (2007) Autologous glioma cell vaccine admixed with interleukin-4 gene transfected fibroblasts in the treatment of patients with malignant gliomas. J Trans Med 5: 67–77.
Onishi Y, Fehervari Z, Yamaguchi T, Sakaguchi S (2008) Foxp3+ natural regulatory T cells preferentially form aggregates on dendritic cells in vitro and actively inhibit their maturation. Proc Natl Acad Sci USA 105: 10113–10118.
Ouaked N, Mantel PY, Bassin C, Burgler S, Siegmund K, Akdis CA, Schmidt-Weber CB (2009) Regulation of the foxp3 gene by the Th1 cytokines: the role of IL-27-induced STAT1. J Immunol 182: 1041–1049.
Oukka M (2008) Th17 cells in immunity and autoimmunity. Ann Rheum Dis 67: 26–29.
Pavelko KD, Heckman KL, Hansen MJ, Pease LR (2008) An effective vaccine strategy protective against antigenically distinct tumor variants. Cancer Res 68: 2471–2478.

Pegram HJ, Jackson JT, Smyth MJ, Kershaw MH, Darcy PK (2008) Adoptive transfer of gene-modified primary NK cells can specifically inhibit tumor progression in vivo. J Immunol 181: 3449–3455.

Perales MA, Yuan J, Powel S, Gallardo HF, Rasalan TS, Gonzalez C, Manukian G, Wang J, Zhang Y, Chapman PB, Krown SE, Livingston PO, Ejadi S, Panageas KS, Engelhorn ME, Terzulli SL, Houghton AN, Wolchok JD (2008) Phase I/II study of GM-CSF DNA as an adjuvant for a multipeptide cancer vaccine in patients with advanced melanoma. Mol Ther 16: 2022–2029.

Perez N, Karumuthil-Melethil S, Li R, Prabhakar BS, Holterman MJ, Vasu C (2008) Preferential costimulation by CD80 results in IL-10-dependent TGF-beta1(+)-adaptive regulatory T cell generation. J Immunol 180: 6566–6576.

Pflanz S, Timans JC, Cheung J, Rosales R, Kanzler H, Gilbert J, Hibbert L, Churakova T, Travis M, Vaisberg E, Blumenschein WM, Mattson JD, Wagner JL, To W, Zurawski S, McClanahan TK, Gorman DM, Bazan JF, de Waal Malefyt R, Rennick D, Kastelein RA (2002) IL-27, a heterodimeric cytokine composed of EBI3 and p28 protein, induces proliferation of naive CD4(+) T cells. Immunity 16: 779–790.

Phares TW, Kean RB, Mikheeva T, Hooper DC 2006. Regional differences in blood-brain barrier permeability changes and inflammation in the apathogenic clearance of virus from the central nervous system. J Immunol 176: 7666–7675.

Piersma SJ, Welters MJ, van der Burg SH (2008) Tumor-specific regulatory T cells in cancer patients. Hum Immunol 69: 241–249.

Pillai V, Ortega SB, Wang CK, Karandikar NJ (2007) Transient regulatory T-cells: a state attained by all activated human T-cells. Clin Immunol 123: 18–29.

Porgador A, Snyder D, Gilboa E (1996) Induction of antitumor immunity using bone marrow-generated dendritic cells. J Immunol 156: 2918–2926.

Powell DJ, Dudley ME, Hogan KA, Wunderlich JR, Rosenberg SA (2006) Adoptive transfer of vaccine-induced peripheral blood mononuclear cells to patients with metastatic melanoma following lymphodepletion. J Immunol 177: 6527–6539.

Prelog M (2006) Aging of the immune system: a risk factor for autoimmunity? Autoimmun Rev 5: 136–139.

Prendergast GC, Jaffee EM (2007) Cancer immunotherapy: immune suppression and tumor growth. Elsevier Publishers, San Diego, CA.

Radhakrishnan S, Cabrera R, Schenk EL, Nava-Parada P, Bell MP, Van Keulen VP, Marler RJ, Felts SJ, Pease LR (2008a) Reprogrammed FoxP3+ T regulatory cells become IL-17+ antigen-specific autoimmune effectors in vitro and in vivo. J Immunol 181: 3137–3147.

Radhakrishnan S, Arneson LN, Upshaw JL, Howe CL, Felts SJ, Colonna M, Leibson PJ, Rodriguez M, Pease LR (2008b) TREM-2 mediated signaling induces antigen uptake and retention in mature myeloid dendritic cells. J Immunol 181: 7863–7872.

Radhakrishnan S, Cabrera R, Bruns KM, Van Keulen VP, Hansen MJ, Felts SJ, Pease LR (2009) Indirect recruitment of a CD40 signaling pathway in dendritic cells by B7-DC cross-linking antibody modulates T cell functions. PLoS One 4: e5373.

Ramoni C, Spadaro F, Menegon M, Podo F (2001) Cellular localization and functional role of phosphatidylcholine-specific phospholipase C in NK cells. J Immunol 167: 2642–2650.

Raulet DH, Guerra N (2009) Oncogenic stress sensed by the immune system: role of natural killer cell receptors. Nat Rev Immunol 9: 568–580.

Rechavi O, Goldstein I, Vernitsky H, Rotblat B, Kloog Y (2007) Intercellular transfer of oncogenic H-Ras at the immunological synapse. PLoS One 2: e1204.

Redman BG, Chang AE, Whitfield J, Esper P, Jiang G, Braun T, Roessler B, Mule JJ (2008) Phase Ib trial assessing autologous, tumor-pulsed dendritic cells as a vaccine administered with or without IL-2 in patients with metastatic melanoma. J Immunother 31: 591–598.

Reich A, Spering C, Schulz JB (2008) Death receptor fas (CD95) signaling in the central nervous system: tuning neuroplasticity? Trends Neurosci 31: 478–486.

Rieger UM, Schlecker C, Pierer G, Haug M (2009) Spontaneous regression of two giant basal cell carcinomas in a single patient after incomplete excision. Tumori 95: 258–263.

Robin MA, Le Roy M, Descatoire V, Pessayre D (1997) Plasma membrane cytochromes P450 as neoantigens and autoimmune targets in drug-induced hepatitis. J Hepatol 26: 23–30.

Robson NC, Hoves, S, Maraskovsky E, Schnurr M (2010) Presentation of tumour antigens by dendritic cells and challenges faced. Curr Opin Immunol. [Epub ahead of print]

Rocha N, Neefjes J (2008) MHC class II molecules on the move for successful antigen presentation. EMBO 27: 1–5.

Rosenberg SA, Packard BS, Aebersold PM, Solomon D, Topalian SL, Toy ST, Simon P, Lotze MT, Yang JC, Seipp CA (1988) Use of tumor-infiltrating lymphocytes and interleukin-2 in the immunotherapy of patients with metastatic melanoma. A preliminary report. N Engl J Med 319: 1676–1680.

Rosenberg SA, White DE (1996) Vitiligo in patients with melanoma: normal tissue antigens can be targets for cancer therapy. J Immunother Emphasis Tumor Immunol 19: 81–84.

Rosenberg SA, Dudley ME (2009) Adoptive cell therapy for the treatment of patients with metastatic melanoma. Curr Opin Immunol 21: 233–240.

Ross D (2003) RNA interference (RNAi): Nuts & Bolts of RNAi Technology. DNA Press, Glendale, CA.

Rothenberg EV (1992) The development of functionally responsive T cells. Adv Immunol 51: 85–214.

Ruggeri L, Capanni M, Urbani E, Perruccio K, Shlomchik WD, Tosti A, Posati S, Rogaia D, Frassoni F, Aversa F, Martelli MF, Velardi A (2002) Effectiveness of donor natural killer cell alloreactivity in mismatched hematopoietic transplants. Science 295: 2097–2100.

Ruppert J, Alexander J, Snoke K (1993) Effect of T cell receptor antagonism on interaction between T-cells and antigen-presenting cells and on T-cell signaling events. Proc Natl Acad Sci USA 90: 2671–2675.

Sakaguchi S, Sakaguchi N, Shimizu J, Yamazaki S, Sakihama T, Itoh M, Kuniyasu Y, Nomura T, Toda M, Takahashi T (2001) Immunologic tolerance maintained by CD25+ CD4+ regulatory T cells: their common role in controlling autoimmunity, tumor immunity, and transplantation tolerance. Immunol Rev 182: 18–32.

Salcedo R, Hixon JA, Stauffer JK, Jalah R, Brooks AD, Khan T, Dai RM, Scheetz L, Lincoln E, Back TC, Powell D, Hurwitz AA, Sayers TJ, Kastelein R, Pavlakis GN, Felber BK, Trinchieri G, Wigginton JM (2009) Immunologic and therapeutic synergy of IL-27 and IL-2: enhancement of T cell sensitization, tumor-specific CTL reactivity and complete regression of disseminated neuroblastoma metastases in the liver and bone marrow. J Immunol 182: 4328–4338.

Samarasinghe R, Tailor P, Tamura T, Kaisho T, Akira S, Ozato K (2006) Induction of an anti-inflammatory cytokine, IL-10, in dendritic cells after toll-like receptor signaling. J Interferon Cytokine Res 26: 893–900.

Schadendorf D, Ugurel S, Schuler-Thurner B, Nestle FO, Enk A, Brocker EB, Grabbe S, Rittgen W, Edler L, Sucker A, Zimpfer-Rechner C, Berger T, Kamarashev J, Burg G, Jonuleit H, Tuttenberg A, Becker JC, Keikavoussi P, Kampgen E, Schuler G (2006) Dacarbazine (DTIC) versus vaccination with autologous peptide-pulsed dendritic cells (DC) in first-line treatment of patients with metastatic melanoma: a randomized phase III trial of the DC study group of the DeCOG. Ann Oncol 17: 563–570.

Schatton T, Frank MH (2009) Antitumor immunity and cancer stem cells. Ann NY Acad Sci 1176: 154–169.

Schena M (1999) DNA Microarrys. Oxford Press, New York, NY.

Schjeide OA, De Vellis J (1970) Cell Differentiation. Van Nostrand Reinhold Co., New York, NY.

Schmittling RJ, Archer GE, Mitchell DA (2008) Detection of humoral response in patients with glioblastoma receiving EGFRvIII-KLH vaccines. J Immunol Methods 339: 74–81.

Schulz TF (2009) Cancer and viral infections in immunocompromised individuals. Int J Cancer 125: 1755–1763.

Screpanti V, Wallin RP, Grandien A, Ljunggren HG (2005) Impact of FASL-induced apoptosis in the elimination of tumor cells by NK cells. Mol Immunol 42: 495–499.

Shimizu J, Yamazaki S, Sakaguchi S (1999) Induction of tumor immunity by removing CD25+CD4+ T cells: a common basis between tumor immunity and autoimmunity. J Immunol 163: 5211–5218.

Silverman GJ, Cary SP, Dwyer DC, Luo L, Wagenknecht R, Curtiss VE (2000) A B cell superantigen-induced persistent "Hole" in the B-1 repertoire. J Exp Med 192: 87–98.

Sinclair NR (2000) Immunoreceptor tyrosine-based inhibitory motifs on activating molecules. Crit Rev Immunol 20: 89–102.

Smith FO, Downey SG, Klapper JA, Yang JC, Sherry RM, Royal RE, Kammula US, Hughes MS, Restifo NP, Levy CL, White DE, Steinberg SM, Rosenberg SA (2008) Treatment of metastatic melanoma using interleukin-2 alone or in conjunction with vaccines. Curr Opin Immunol 21: 233–240.

Sojka DK, Huang YH, Fowell DJ (2008) Mechanisms of regulatory T-cell suppression—a diverse arsenal for a moving target. Immunology 124: 13–22.

Sprent J (1999) Presidential address to the American Association of Immunologists: Naïve T cells. J Immunol 163: 4629–4636.

Stackstein R, Janssen WE, Elfenbein GJ (1995) Bone Marrow Transplantation: Foundations for the 21st Century. New York Academy of Sciences, NY.

Stout RD, Watkins SK, Suttles J (2009) Functional plasticity of macrophages: in situ reprogramming of tumor-associated macrophages. J Leukoc Biol 86: 1105–1109.

Strowig T, Brilot F, Munz C (2008) Noncytotoxic functions of NK cells: direct pathogen restriction and assistance to adaptive immunity. J Immunol 180: 7785–7791.

Sun JC, Beilke JN, Lanier LL (2009a) Adaptive immune features of natural killer cells. Nature 457: 557–561.

Sun Y, Chin YE, Weisiger E, Malter C, Tawara I, Toubai T, Gatza E, Mascagni P, Dinarello CA, Reddy P (2009b) Cutting edge: Negative regulation of dendritic cells through acetylation of the nonhistone protein STAT-3. J Immunol 182: 5899–5903.

Sutlu T, Alici E (2009) Natural killer cell-based immunotherapy in cancer: current insights and future prospects. J Intern Med 266: 154–181.

Sutmuller RP, van Duivenvoorde LM, van Elsas A, Schumacher TN, Wildenberg ME, Allison JP, Toes RE, Offringa R, Melief CJ (2001) Synergism of cytotoxic T lymphocyte-associated antigen 4 blockade and depletion of CD25(+) regulatory T cells in antitumor therapy reveals alternative pathways for suppression of autoreactive cytotoxic T lymphocyte responses. J Exp Med 194: 823–832.

Szabo SJ, Kim ST, Costa GL, Zhang X, Fathman CG, Glimcher LH (2000) A novel transcription factor, T-bet, directs Th1 lineage commitment. Cell 100: 655–669.

Taams LS, van Amelsfort JM, Tiemessen MM, Jacobs KM, de Jong EC, Akbar AN, Bijlsma JW, Lafeber FP (2005) Modulation of monocyte/macrophage function by human CD4+CD25+ regulatory T cells. Hum Immunol 66: 222–230.

Tang J, Flomenberg P, Harshyne L (2005) Glioblastoma patients exhibit circulating tumor-specific CD8 T cells. Clin Cancer Res 11: 5292–5299.

Terabe M, Berzofsky JA (2008) The role of NKT cells in tumor immunity. Adv Cancer Res 101: 277–348.

te Velde AA, Figdor CG (1992) Monocyte mediated cytotoxic activity against melanoma. Melanoma Res 1: 303–309.

Tiemessen MM, Jagger AL, Evans HG, van Herwijnen MJ, John S, Taams LS (2007) CD4+CD25+Foxp3+ regulatory T cells induce alternative activation of human monocytes/macrophages. Proc Natl Acad Sci USA 104: 19446–19451.

Turkington C, LiPera WL (2005) Encyclopedia of Cancer. Facts On File, Inc., New York, NY.

Turner SJ, La Gruta NL Kedzierska K, Thomas PG, Doherty PC (2009) Functional implications of T cell receptor diversity. Curr Opin Immunol 3:286–290.

Valone FH, Kaufman PA, Guyre PM Lewis LD, Memoli V, Deo, Y, Graziano R, Fisher JL, Meyer L, Mrozek-Orlowski M (1995) Phase Ia/Ib trial of bispecific antibody MDX-210 in patients with advanced breast or ovarian cancer that overexpresses the proto-oncogene HER-2/neu. J Clin Oncol 13: 2281–2292.

Valitutti S, Dupre L (2010) Plasticity of immunological synapses. Curr Top Microbiol Immunol 340: 209–228.

Vandewoestijne S, Schtickzelle N, Baguette M (2008) Positive correlation between genetic diversity and fitness in a large, well-connected metapopulation. BMC Biol 6: 46–55.

Velardi A, Ruggeri L, Mancusi A, Aversa F, Christiansen FT (2009) Natural killer cell allorecognition of missing self in allogeneic hematopoietic transplantation: a tool for immunotherapy of leukemia. Curr Opin Immunol 21: 525–530.

Vieweg J, Su Z, Dahm P, Kusmartsev S (2007) Reversal of tumor-mediated immunosuppression. Clin Cancer Res 13: 727–732.

Viguier M, Lemaitre F, Verola O, Cho MS, Gorochov G, Dubertret L, Bachelez H, Kourilsky P, Ferradini L (2004) Foxp3 expressing CD4+CD25(high) regulatory T cells are overrepresented in human metastatic melanoma lymph nodes and inhibit the function of infiltrating T cells. J Immunol 173: 1444–1453.

Visovsky CG (2005) Peripheral nerve function and symptom distress during biotherapy for malignant melanoma. Can Oncol Nurs 15: 161–174.

Vogl TJ, Naguib NN, Nour-Eldin NE, Eichler K, Zangos S, Gruber-Rouh T (2010) Transarterial chemoembolization (TACE) with mitomycin C and gemcitabine for liver metastases in breast cancer. Eur Radiol 20: 173–180.

Volpato M, Phillips RM (2007) Tailoring targeted therapy to individual patients: lessons to be learnt from the development of mitomycin C. Cancer Genomics Proteomics 4: 175–186.

von Euw E, Chodon T, Attar N, Jalil J, Koya RC, Comin-Anduix B, Ribas A (2009) CTLA4 blockade increases Th17 cells in patients with metastatic melanoma. J Transl Med 7: 35–44.

Walker MR, Kasprowicz DJ, Gersuk VH, Benard A, Van Landeghen M, Buckner JH, Ziegler SF (2003) Induction of FoxP3 and acquisition of T regulatory activity by stimulated human CD4+CD25- T cells. J Clin Invest 112: 1437–1443.

Walker MR, Carson BD, Nepom GT, Ziegler SF, Buckner JH (2005) De novo generation of antigen-specific CD4+CD25+ regulatory T cells from human CD4+CD25-cells. Proc Natl Acad Sci USA 102: 4103–4108.

Walther W, Ulrike S (2000) Gene Therapy of Cancer: Methods and Protocols. Humana Press (Springer), New York, NY.

Wang HY, Peng G, Guo Z, Shevach EM, Wang RF (2005) Recognition of a new ARTC1 peptide ligand uniquely expressed in tumor cells by antigen-specific CD4+ regulatory T cells. J Immunol 174: 2661–2670.

Wang T, Niu G, Kortylewski M, Burdelya L, Shain K, Zhang S, Bhattacharya R, Gabrilovich D, Heller R, Coppola D, Dalton W, Jove R, Pardoll D, Yu H (2004) Regulation of the innate and adaptive immune responses by Stat-3 signaling in tumor cells. Nat Med 10: 48–54.

Warrell DA (2005) Oxford Textbook of Medicine: Sections 18–33, volume 3. Oxford University Press, Oxford.

Wei J, Barr J, Kong LY, Wang Y, Wu A, Sharma AK, Gumin J, Henry V, Colman H, Sawaya R, Lang FF, Heimberger AB (2010) Glioma-associated cancer-initiating cells induce immunosuppression. Clin Cancer Res 16: 461–473.

Wei WZ, Jacob J, Radkevich-Brown O, Whittington P, Kong YC (2008) The "A, B and C" of her-2 DNA vaccine development. Cancer Immunol Immunother 57: 1711–1717.

Westcott P (1999) Living with Leukemia. Heinmann-Raintree, Chicago, IL. Wierda WG, Kipps TJ (2007) Gene therapy and active immune therapy of hematologic malignancies. Best Pract Res Clin Haematol 20: 557–568.

Wheeler CJ (2010) Dendritic cell vaccines to combat glioblastoma. Expert Rev Neurother 10: 483–486.

Winstead CJ, Reilly CS, Moon JJ, Jenkins MK, Hamilton SE, Jameson SC, Way SS, Khoruts A (2010) CD4+CD25+Foxp3+ regulatory T cells optimize diversity of the conventional T cell repertoire during reconstitution from lymphopenia. J Immunol. [Published online March 31, 2010]

Wierda WG, Kipps TJ (2007) Gene therapy and active immune therapy of hematologic malignancies. Best Pract Res Clin Haematol 3: 557–568.

Wolchok JD, Saenger Y (2008) The mechanism of anti-CTLA-4 activity and the negative regulation of T-cell activation. Oncologist 13: 2–9.

Wolfl M, Rutebemberwa A, Mosbruger T, Mao Q, Li HM, Netski D, Ray SC, Pardoll D, Sidney J, Sette A, Allen T, Kuntzen T, Kavanagh DG, Kuball J, Greenberg PD, Cox AL (2008) Hepatitis C virus immune escape via exploitation of a hole in the T cell repertoire. J Immunol 181: 6435–6446.

Wright KL, Ting JP (2006) Epigenetic regulation of MHC-II and CIITA genes. Trends Immunol 27: 405–412.

Yamazaki S, Dudziak D, Heidkamp GF, Fiorese C, Bonito AJ, Inaba K, Nussenzweig MC, Steinman RM (2008) CD8+ CD205+ splenic dendritic cells are specialized to induce Foxp3+ regulatory T cells. J Immunol 181: 6923–6933.

Yamazaki S, Iyoda T, Tarbell K, Olson K, Velinzon K, Inaba K, Steinman RM (2003) Direct expansion of functional CD25+ CD4+ regulatory T cells by antigen-processing dendritic cells. J Exp Med 198: 235-247.

Yi AK, Yoon JG, Yeo SJ, Hong SC, English BK, Krieg AM (2002) Role of mitogen-activated protein kinases in CpG DNA-mediated IL-10 and IL-12 production: central role of extracellular signal-regulated kinase in the negative feedback loop of the CpG DNA-mediated Th1 response. J Immunol 168: 4711–4720.

Yigit R, Massuger LF, Figdor CG Torensma R (2010) Ovarian cancer creates a suppressive microenvironment to escape immune elimination. Gynecol Oncol (Epub ahead of print).

Zhang X, Jin J, Tang Y, Speer D, Sujkowska D, Markovic-Plese S (2009) IFN-beta1a inhibits the secretion of Th17-polarizing cytokines in human dendritic cells via TLR7 up-regulation. J Immunol 182: 3928–3936.

Zhao Y, Zheng Z, Robbins PF, Khong HT, Rosenberg SA, Morgan RA (2005) Primary human lymphocytes transduced with NY-ESO-1 antigen-specific TCR genes recognize and kill diverse human tumor cell lines. J Immunol 174: 4415–4423.

Ziegler SF (2007) FOXP3: not just for regulatory T cells anymore. Eur J Immunol 37: 21–23.

Index

G

H

I

L

M

N

P